AF333032

World Soils Book Series

Series editor

Prof. Alfred E. Hartemink
Department of Soil Science, FD Hole Soils Laboratory
University of Wisconsin–Madison
Madison
USA

For further volumes:
http://www.springer.com/series/8915

Aims and Scope

The *World Soils Book Series* brings together soil information and soil knowledge of a particular country in a concise and reader-friendly way. The books include sections on soil research history, geomorphology, major soil types, soil maps, soil properties, soil classification, soil fertility, land use and vegetation, soil management, and soils and humans.

International Union of Soil Sciences

Zueng-Sang Chen · Zeng-Yei Hseu
Chen-Chi Tsai

The Soils of Taiwan

Springer

Zueng-Sang Chen
Department of Agricultural Chemistry
National Taiwan University
Taipei
Taiwan

Chen-Chi Tsai
Department of Forestry and Natural Resources
National Ilan University
Ilan
Taiwan

Zeng-Yei Hseu
Department of Environmental Science
  and Engineering
National Pingtung University of Science
  and Technology
Pingtung
Taiwan

ISSN 2211-1255
World Soils Book Series
ISBN 978-94-017-9725-2
DOI 10.1007/978-94-017-9726-9

ISSN 2211-1263   (electronic)

ISBN 978-94-017-9726-9   (eBook)

Library of Congress Control Number: 2014959263

Springer Dordrecht Heidelberg New York London

Printed on acid-free paper

Springer Science+Business Media B.V. Dordrecht is part of Springer Science+Business Media (www.springer.com)

Preface

Formosa, the original name of Taiwan, was given to the island in 1535 by Portuguese sailors who spotted the main island of Taiwan, and named it "Ilha Formossa" which means beautiful island, a fitting epithet for an island with a great natural beauty. Taiwan is an island where a high variation in climate conditions produces a great biodiversity of vegetation types while, at the same time, it is situated on a complex convergent boundary between the Euroasian and Pacific plate which causes a spectacular landscape with amazing mountain ranges, one of the highest elevations of Taiwan being near 4,000 m, deep gorges, and breathtaking valleys. Taiwan has almost all the major soil types of the world except permafrost soils. In other words, Taiwan is a beautiful open air museum of world soils.

Taiwan's modern soil research was started by the Japanese in the 1920s, who published a soil texture and soil pH map for soil management on crop production. The real soil survey project was under the auspices of American experts, while the soil classification system of the U.S.A. was used to conduct the reconnaissance and detailed soil survey of Taiwan which started in 1946. The more soil survey projects were undertaken, the more soil information about rural soils, hill lands, and high mountain forest soils of Taiwan was provided in order to satisfy the need for food security, economic growth, and environmental quality in the last 4–5 decades. The members of the Chinese Society of Soil and Fertilizer Sciences also continued their contributing research on Taiwan soils. Nowadays, climate change and industrial development form a great challenge to the prevention of soil degradation, the remediation of already contaminated soil, as well as retaining the protection and continuity of food safety. This book can provide the most typical soil characteristics as well as basic soil information through soil survey reports and the new soil information system of Taiwan.

Land uses and land management in Taiwan are major influential factors in relation to Taiwan's economic growth, industrial development, and the degradation of the environmental quality of soil, especially in the 1970s–1990s. The rural soils were contaminated by the illegally discharged wastewater from industrial parks, while hill lands were reshaped by the production of high economic fruit crops both of which caused serious soil erosion and soil degradation during the last two decades. Therefore the protection of soil resources, including soil and water conservation, remediation of contaminated soils, and the conservation of high mountain forest regions must become more and more important in the future.

In the last decade, the development of soil quality monitoring projects and a soil information system combined together in a geographical information system have become increasingly important since, as a result, real-time database is able to provide vital information to predict where highly contaminated soils are likely to be situated, thus preventing the production of food crops that could endanger the health and, at the same time, maintaining the food safety. This is why, different governmental offices in Taiwan share their data of air quality, irrigation water, soil, groundwater, and the location of industrial plants with soil researchers and communities in Taiwan.

Therefore, studies in Soil Science are more important than ever before. The government must also continue to provide scholarships and funds to attract more students to study topics like soil amendment, soil remediation, and soil degradation in the future. At the same time, the exhibition of soil monoliths for farmers and a regional center and soil museum for both students as well as communities have been established in Taiwan since 2000. Still, for all these activities we need strong governmental support to be able to retain Taiwan's beauty with all its varied soils, landscapes, flora, and fauna.

Acknowledgments

The authors thank Drs./Profs. Angelier, Chen et al., Chiang, Hseu et al., Wen-Hsu Huang, Shih-Hao Jien, Lee, Maydell, Tsai, and Wang et al. for preparing making Figures and Tables.

Contents

Authors' Biography

Zueng-Sang Chen is Distinguished Professor of pedology and soil environmental quality (2007 to now), Department of Agricultural Chemistry (DAC) of National Taiwan University (NTU). He was the Associate Dean of College of Bioresources and Agriculture of NTU in 2007–2011 and Department Head of DAC/NTU in 2004–2007. He was awarded the Distinguished Agricultural Expert Award of Council of Agriculture of Taiwan in 2012, the East and Southeastern Federation of Soil Science Societies (ESAFS) Distinguished Award in 2009, NTU Distinguished Social Service Award in 2009, and the KIWANIS Distinguished Agricultural Expert Award in 2007. He primarily studied soil genesis, soil environmental quality, the behavior and bioavailability of heavy metals in the soil–crop system, and using phytoremediation on metals-contaminated sites.

Zeng-Yei Hseu works as Full Professor at Department of Environmental Science and Engineering, National Pingtung University of Science and Technology (NPUST), Taiwan. He also worked as the Chairman of this department from 2013 to 2016. His major topics of interest include heavy metal dynamics and mineralogy of serpentine soil, morphology and genesis of wetland soil, soil chronosequences on river and marine terraces, and soil heavy metal contamination and remediation. He is the author or coauthor of about 90 scientific papers and book chapters.

Chen-Chi Tsai is Full Professor of Pedology at Department of Forestry and Natural Resources, National Ilan University (NIU). His main research interests include soil classification and pedogenesis of volcanic soils and forest soils, carbon sequestration techniques, and pool of forest soils with emphasis on the different land use, vegetation type, elevation and climate condition, and effects of biochar application on soil physiochemical properties and plant growth.

1.1 Introduction

Taiwan is an island located about 150 km off the southeast coast of Mainland China. It covers 36,000 km^2 in area with a population of more than 23 million residents. The Central Mountain Range is the most prominent feature in landscape (Fig. 1.1) (Lin 1957; Ho 1988). There are more than 200 peaks over 3,000 m in altitude, whose geographical environment supports a rich fauna, flora, and biological diversity. There are over 4,000 vascular plant species and a spectrum of six forest types in the area. The orogenesis of these high mountains has resulted from the arc-continent collision between the Eurasian plate and Philippine Sea plate since the Plio-Pleistocene (Fig. 1.2). The landscape decreases westward and eastward in altitude from the north-south trending ridge. In addition to the mountainous area, the hills and tablelands account for 40 %, and plains for the remaining 30 % of the total area. The coastlines along the island are characterized types of eastern fault coast, western emergent coast, northern mixed coast, and southern coral-reef coast (Lin 1957). The Tropic of Cancer (23.5° N) running across Taiwan's middle section divides the island into two climactic zones, tropical in the south and subtropical in the north. The island's average annual temperature is about 24 °C in the south and 22 °C in the north. The main stream of the northward-moving Kuroshio Current passes up the eastern coast of Taiwan, thus bringing in warm and moist air. Summer and winter monsoons also bring intermittent rainfall to Taiwan's hills and central mountains. In general, there is more than 2,500 mm of rainfall every year. The complex environment in Taiwan develops a wide variety of soils on the surfaces over the island. The soils in Taiwan consist of 11 of the 12 soil orders in *Soil Taxonomy* with the exception of Gelisol. The various landforms and vegetation provide an excellent and convenient opportunity for studies on the soil forming process.

1.2 Geomorphology and Geology

Taiwan is located at the convergent boundary between the Philippine Sea plate and the continental margin of the Eurasian plate (Fig. 1.2a). The Philippine Sea plate subducts beneath the Eurasian plate offshore eastern Taiwan, and override the South China Sea floor of the Eurasian plate south of Taiwan (Suppe 1981; Tsai 1986; Angelier 1986; Ho 1988) (Fig. 1.2b). The collision by oblique plate convergence began at 5–6 Ma, and is ongoing with continual crustal shortening and widespread seismic activities (Ho 1986; Teng 1990; Seno et al. 1993). Recent GPS data indicate a crustal shortening by 8.5 cm/year across the 200 km-wide island (Yu et al. 1997). According to the lithology and tectonic environment of Taiwan, the active tectonic results in five geologic provinces, which are Coastal Plain, Western Foothills, Central Range, Longitudinal Valley, and Coastal Range (Ho 1988) (Fig. 1.3a).

The 10−20 km-wide Western Foothills of Taiwan are composed of thin-skinned fold and thrust belt that flank the Central Range along much of its length. This fold and thrust belt were developed by sequential thrusting of a series of subparallel reverse faults westward by tectonics compression during the Plio-Pleistocene (Lee et al. 1996) (Fig. 1.3b). The strong geotectonic stress is currently recorded in central Taiwan. For instance, the largest inland earthquake ($M_w = 7.6$, $M_L = 7.3$) of the twentieth century in Taiwan was caused by the displacement of the Chelungpu Fault in this provinces (Kao and Chen 2000; Ma et al. 1999). However, northern Taiwan is currently experiencing a post collision stage because of southward migration of the collision zone since 5−6 Ma. In addition, back-arc rifting is occurring within the Ryukyu Arc system, whose geomorphic expression is the Okinawa Trough. This rifting has gradually migrated into northern Taiwan, as collision-related compression has largely ceased in the area (Lu et al. 1995; Teng 1996).

Z.-S. Chen et al., *The Soils of Taiwan*, World Soils Book Series,
DOI 10.1007/978-94-017-9726-9_1, © Springer Science+Business Media Dordrecht 2015

Fig. 1.1 The satellite image of Taiwan. Copyright by Dr. Wen-Hsu Huang

Regarding the geological compositions, the major rock formations of Taiwan form long narrow belts roughly parallel to the long axis of the island (Fig. 1.4). These rock belts young progressively westward from the central backbone range to the west coast. The prevailing structural pattern is that of an elongate arc convex to the west. The northern, shorter bend of the arc strikes east-northeast whereas the southern, major arm of the arc extends mainly north-south. All the major structures of Taiwan, including the important faults and fold axes, follow fairly closely this arcuate structure throughout the island.

The metamorphic basement of Taiwan is the oldest geologic-tectonic element, and was formed in Late Paleozoic to Mesozoic times when a thick sequence of sandstone,

Fig. 1.2 The tectonic setting of **a** Asia, and the **b** 3D model for Taiwan. Copyright by Angelier (1986)

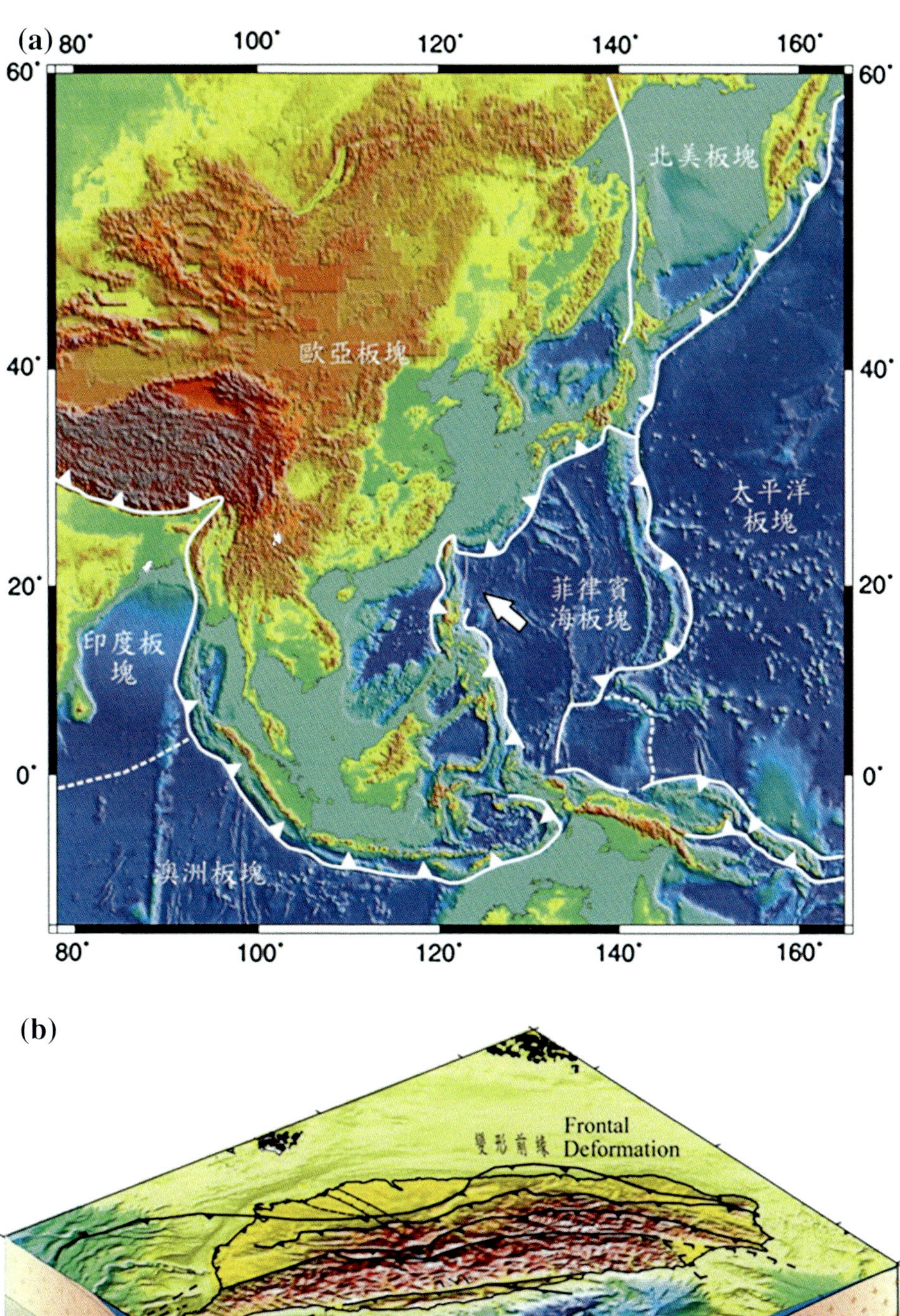

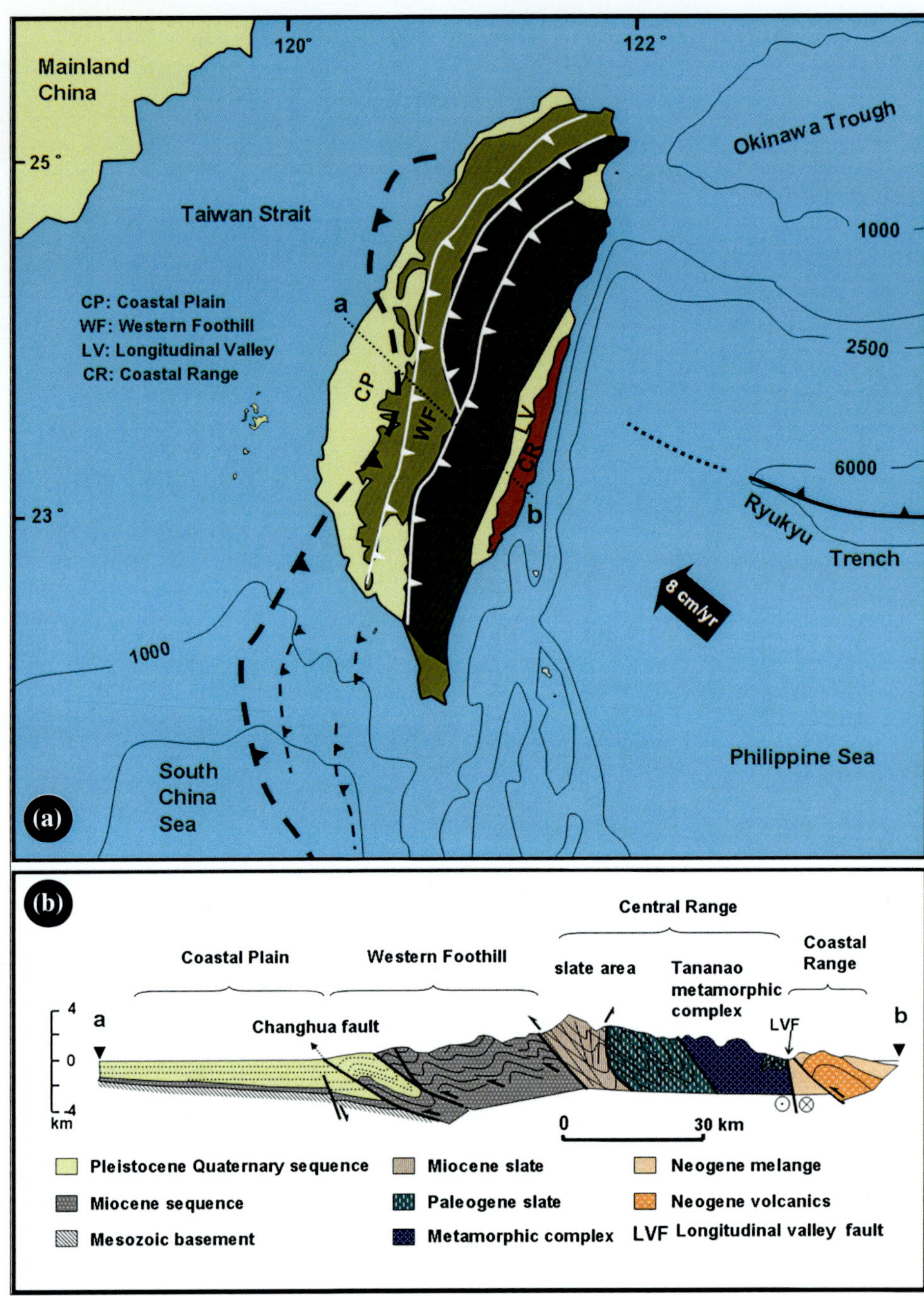

Fig. 1.3 **a** Five geologic provinces of Taiwan and **b** the cross section along AB in (**a**). Copyright by Dr. Wen-Hsu Huang

shale, siltstone, limestone, and volcanic rocks was deposited along with accompanying acid to intermediate magmatic activity. This old geologic-tectonic element experienced several phases of orogenic deformation, magmatism, and metamorphism, culminating in a Late Mesozoic orogeny, locally known as the Nanao Orogeny. These pre-Cenozoic rocks were then folded into mountain ranges and intensely metamorphosed to form the main metamorphic complex of Taiwan. The detailed geologic history of this metamorphic basement remains difficult to unravel. For the consideration of soil parent materials, the metamorphic rocks are mainly exposed on the eastern parts of Taiwan, sedimentary rocks are on western Taiwan, and a few volcanic rocks and their volcanic ash are on northern Taiwan.

Fig. 1.4 Geological map of Taiwan. Copyright by Dr. Wen-Hsu Huang

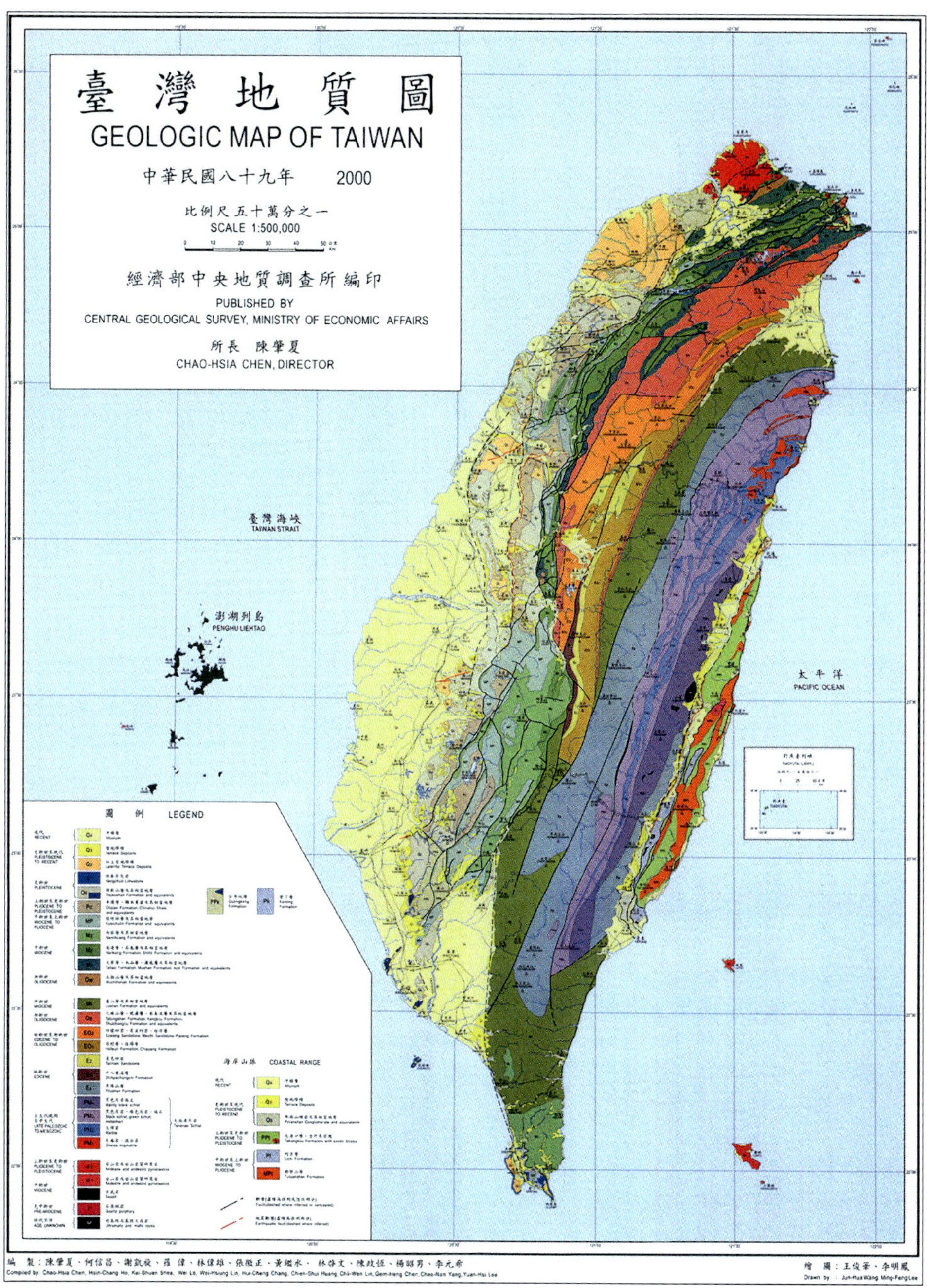

1.3 Climate and Vegetation

The Tropic of Cancer (23.5° N) runs across Taiwan and divides the island into two climactic zones, tropical in the southern and western plains and subtropical in the northern and mountainous regions. The island's average annual temperature is about 24 °C in the south and 22 °C in the north. There are five climatic regions in the main island of Taiwan, according to the Koppen Classification System (Chen 1986) (Fig. 1.5). In general, the climate is characterized by high temperature and humidity, substantial rainfall, and tropical cyclones in summer, alternated with the cool and dry condition of winter. The warmest average

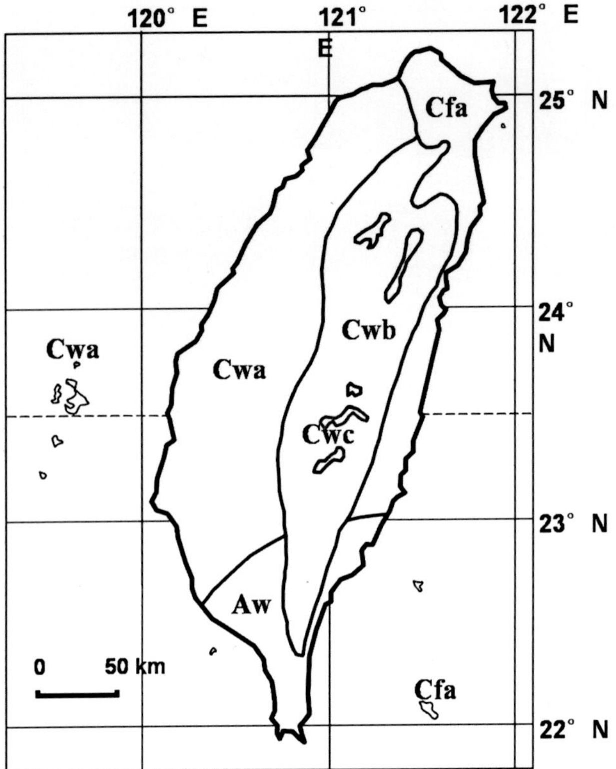

Fig. 1.5 The climatic types of Taiwan based on Koppen classification. Copyright by Dr. Wen-Hsu Huang

temperature is in July, whereas the coldest average temperature is in January. The dense forests covering more than 60 % of Taiwan resulted from the abundant rainfall brought by typhoon, monsoons, trade wind, and airstreams.

The paleoclimate of Taiwan is different from the climate observed in present-day. It was characterized by glacial-interglacial, warm-cold-warm cycles in Pleistocene. The cold and dry mid-Pleistocene phase is considered to be younger than the Jaramillo event (about 0.7–0.9 Ma) (Tseng et al. 1992; Liew and Huang 1994). Palynological evidence indicates a 1–2 °C warmer mean annual temperature, and higher precipitation during the warm interval than during the mid-Pleistocene cool-dry interval (Liew and Hsieh 2000).

Since that time, the warm-temperate to subtropical climatic conditions continued and were interrupted by three early and mid-Holocene very humid intervals (Chen and Liew 1990; Liew and Hsieh 2000). Huang et al. (1997) show that the East Asian monsoon system fluctuated significantly from a strengthened winter monsoon during the most recent glaciation (12–25 ka), through moderate to weak winter and summer monsoons during deglaciation (10–12 ka), to an enhanced summer monsoon in the Holocene.

The current climate of Taiwan is characterized by high temperature and humidity. Heavy rainfall and typhoons (tropical cyclones) affect the island frequently in summer. The mean annual temperature measures 25 °C, within an averaged range between 15 and 30 °C throughout northern and central to southern Taiwan. The annual rainfall, ranging 1,800–2,400 mm, is mainly precipitated from June to September (Fig. 1.6). The northeast monsoon winds follow and prevail till March of the next year. The seasonal variation is between dry in winter and wet in summer.

1.4   Human Activity Factors

Human-induced processes that markedly change soil properties and result in diagnostic horizons or properties are anthropogenic processes. In the case of soils affected by human activity in large-scale farming, the anthropogenic processes have been created in Taiwan for approximately 400 years, particularly based on numerous agricultural activities in soils, such as puddling surface soils (mechanically stirring and mixing surface soil with water and making it into a muddy paste) for paddy rice production by humans. Four hundred years ago, the migrants from China began to clearly reclaim the land of Taiwan. Before this period, a few aborigines were the major people in this island, for whom hunting and fishing were the major living style and thus the soils had trace impacts of humans. The migrants brought farming techniques into the island, so that the forests became crop lands particularly in the western part of Taiwan including alluvial plains and mountains with low elevation.

Fig. 1.6 The averaged monthly temperature and precipitation of Taiwan: **a** Northern Taiwan (Taoyuan); **b** Central Taiwan (Taichung); **c** Southern Taiwan (Chiayi). Copyright by Dr. Wen-Hsu Huang

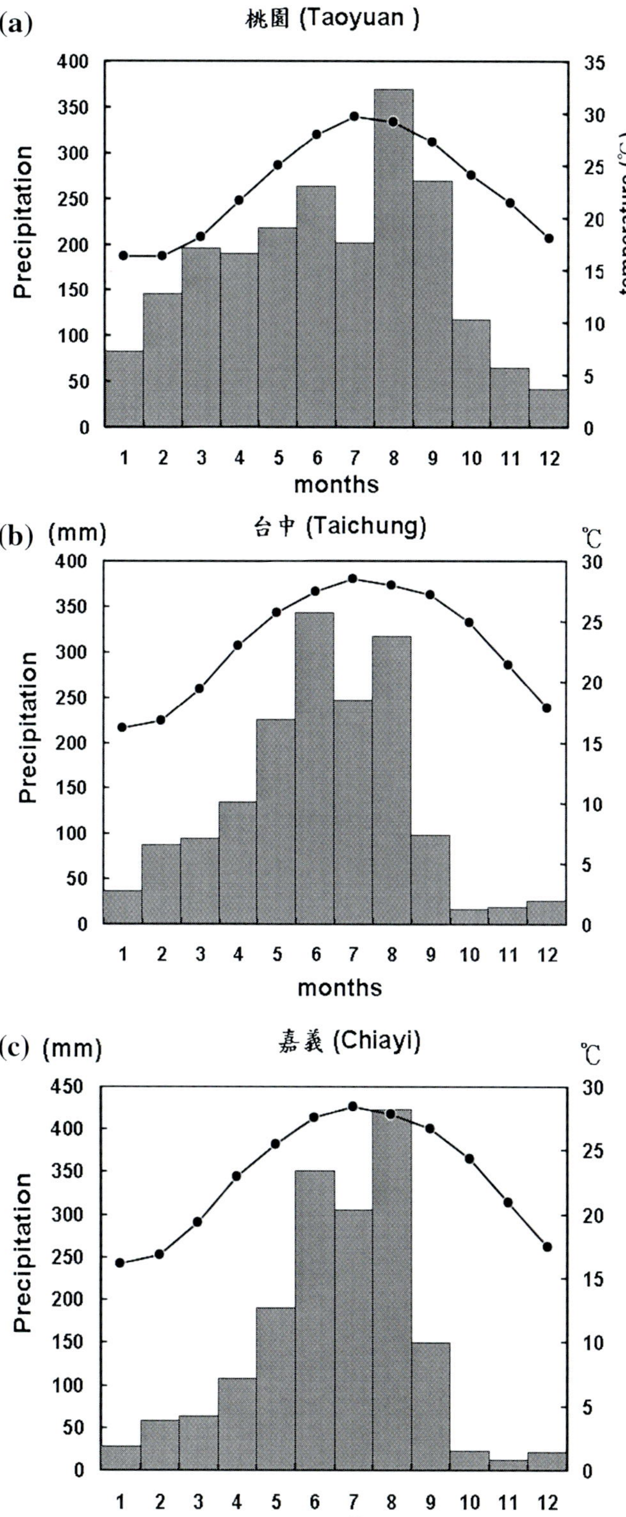

Because paddy rice is the major staple food of the Chinese, the crop lands were dominated by paddy fields. Therefore, frequent submergence and plowing for rice production is the major factor in soil development by human activity in Taiwan since the eighteenth century. With increasing population and the discovery of mechanical power, the area of paddy soils expanded, particularly since the mid-twentieth century in Taiwan. In addition to paddy, farming of sugarcane and orchards was frequent after the nineteenth century.

Paddy soils are developed from various other soils. Land leveling and terracing produce changes in soil moisture regime for paddy rice production. The recognition of paddy soils is mainly based on the obvious impacts of anthropogenic activities on the soils. Artificial and seasonal water saturation brings about enhanced reduction, illuviation, and oxidation processes of iron (Fe) and manganese (Mn) in the soil profile, often leading to the formation of special layers characterized by Fe and Mn distribution. Puddling is often a necessary practice for easy seedling transplantation, which results in a poor soil structure in the plow layer and a compacted plow pan, the latter being good for saving irrigation water so long as it does not adversely affect the root growth. Therefore, a principal distinction of paddy soils from upland soils is the variation in oxidation and reduction status.

1.5 Soil Classification

A total of 620 soil series in Taiwan rural soils can be classified as nine soil orders based on Soil Taxonomy (Table 1.1 and Fig. 1.7). The major soil orders distributed in the rural soils are Inceptisols (51.0 %), Alfisols (21.8 %), Ultisols (9.6 %), and Entisols (6.8 %). The other soil orders occupy only <1 % of the total rural soils. Nine soil orders were also classified in Taiwan forest soils (Table 1.2). The major soil orders distributed in the forest soils are Inceptisols (44.2 %), Entisols (35.3 %), Alfisols (10.8 %), and Ultisols (7.50 %). The other soil orders occupy only <1 % of the total forest soils in Taiwan.

Table 1.1 The distribution of soil orders in Taiwan cultivated soils[a]

Soil orders	Approximate area (km^2)	Percentage of land use	Major soil taxa of great group[b]
Inceptisols	8,590	51.0	Endoaquepts, Epiaquepts, Dystrochrepts, Haplochrepts
Alfisols	3,668	21.8	Hapludalfs, Haplustalfs, Paleustalfs
Ultisols	1,624	9.6	Paleaquults, Rhodudults, Hapludults, Paleudults
Entisols	1,142	6.8	Udorthents, Udipsamments, Udifluents, Haplaquents
Andisols	195	1.2	Hapludands
Mollisols	191	1.1	Hapludolls
Oxisols	50	0.3	Hapludox, Haplustox
Histosols	6	0.04	Medihemists
Vertisols	1	<0.01	Hapluderts
Miscellaneous lands	1,363	8.1	–
Subtotal	16,830	100.0	

[a] Modified from Wang et al. (1988) and Chen (1992)
[b] Soil Taxonomy (Soil Survey Staff 1994)
Chen and Hseu (1997)

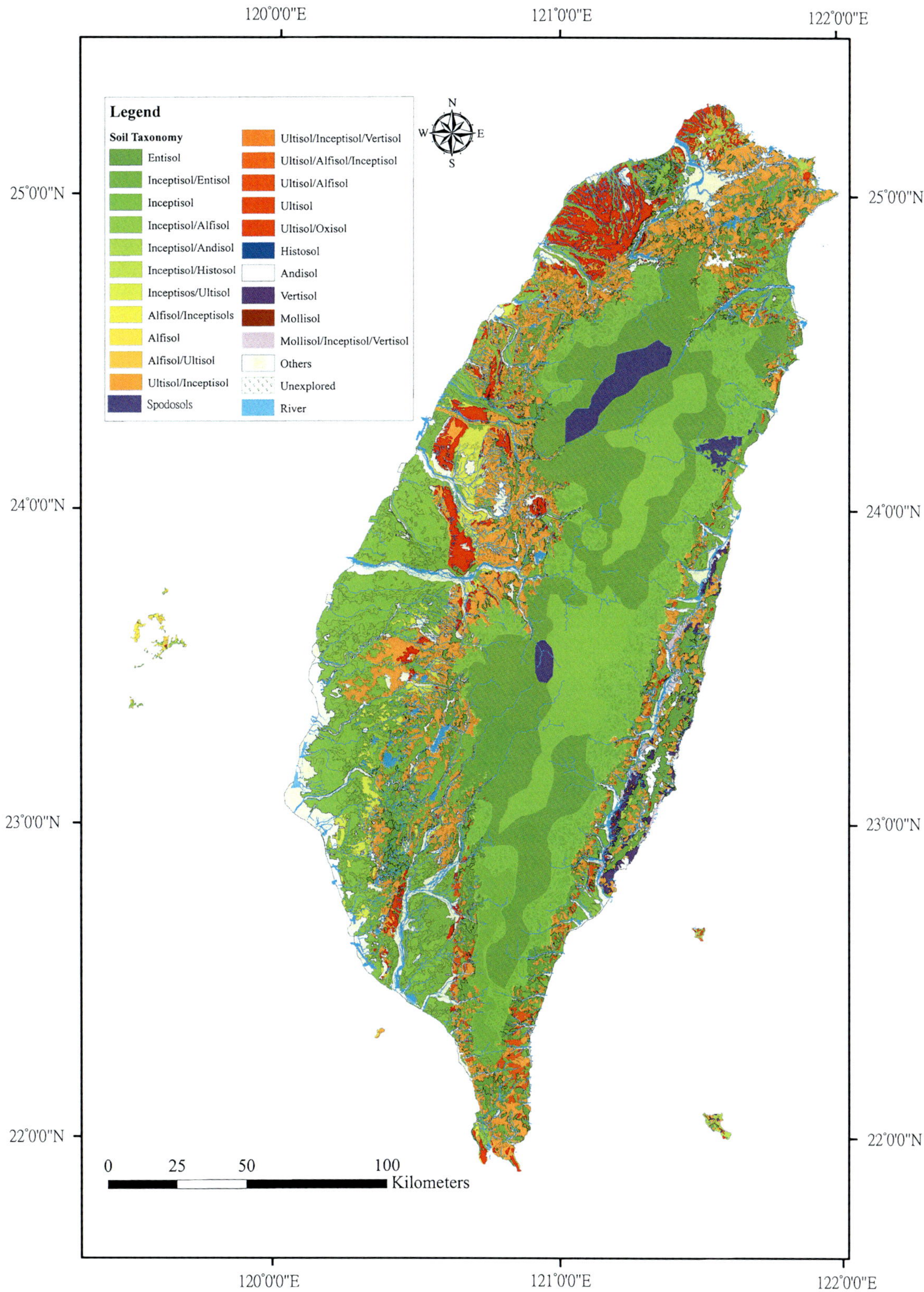

Fig. 1.7 Soil order map of Taiwan. Copyright by Dr. Shih-Hao Jien

Table 1.2 The distribution of soil orders in Taiwan forest soils[a]

Soil orders	Approximate area (km^2)	Percentage of land use	Major oil taxa of great group[b]
Inceptisols	7,415	44.2	Dystrochrepts, Haplochrepts
Entisols	5,923	35.3	Udorthents, Udarents
Alfisols	1,805	10.8	Hapludalfs
Ultisols	1,250	7.50	Hapludults, Paleudults
Andisols	184	1.10	Hapludands, Melaudands
Mollisols	120	0.72	Hapludolls
Spodosols	10	0.06	Haplorthods
Histosols	2	0.01	Medihemists
Vertisols	1	<0.01	Hapluderts
Miscellaneous lands	60	0.36	–
Subtotal	16,770	100.0	

[a] Modified from Chen and Chiang (1996)
[b] Soil Taxonomy (Soil Survey Staff 1994)
(Chen and Hseu 1997)

References

Angelier J (1986) Geodynamics of eurasia-philippine sea plate boundary. Preface Tectonophys 125:IX–X

Chen FY, Liew PM (1990) Palynological study of the Sungshan formation, Taipei basin. Proc Geol Soc China 33:21–37

Chen KY (1986) Koppen's classification on the climatic probability of Taiwan. Geogra. Res. 12:45–56.

Chen ZS (1992) Morphological characteristics, pedogenic processes, and classification of wet soils in Taiwan. In: Kimble JM (ed) Proceeding of the eighth international soil classification Workshop. Baton Rouge, Houston, and San Antonio, pp 53–59, 6–21 Oct 1990

Chen ZS, Chiang HC (1996) Soil characteristics, genesis and classification of some forest soils in Taiwan. In: Proceedings of international congress on soils of tropical forest ecosystem, (3rd Congress on Forest Soils), (ISSS-AISS-IBG), vol 1: 63–78

Chen ZS, Hseu ZY (1997) Total organic carbon pool in soils of Taiwan. Proc Natl Sci Counc ROC B: Life Sci 21:120–127

Ho CS (1986) A synthesis of the geologic evolution of Taiwan. Tectonophysics 125:1–16

Ho CS (1988) An introduction to the geology of Taiwan: explanatory text of the geologic map of Taiwan, 2nd edn. Central Geologic Survey, Taiwan, Republic of China

Huang CY, Liew PM, Zhao M, Chang TC, Kuo CM, Chen MT, Wang CH, Zheng LF (1997) Deep sea and lake records of the Southeast Asian paleomonsoons for the last 25 thousand years. Earth Planet Sci Lett 146:59–72

Kao H, Chen WP (2000) The chi-chi earthquake sequence: active out-of-sequence thrust faulting in Taiwan. Sci 288:2346–2349

Lee JC, Lu CY, Chu HT, Delcaillau B, Angelier J, Deffontaines B (1996) Active deformation and paleostress analysis in the Pakua anticline area, western Taiwan. Terr Atmos Oceanic Sci 7:431–446

Liew PM, Hsieh ML (2000) Late Holocene (2 ka) sea level, river discharge and climate interrelationship in the Taiwan region. J Asian Earth Sci 18:499–505

Liew PM, Huang SY (1994) Pollen analysis and their paleoclimatic implication in the middle Pleistocene lake deposits of the Ilan district, northeastern Taiwan. J Geol Soc China 37:115–124

Lin CC (1957) Topography of taiwan: publication of the Taiwan provincial documentary committee, 424 pp [in Chinese]

Lu CY, Angelier J, Chu HT, Lee JC (1995) Contraction, transcurrent, rotational and extensional tectonics: examples from Northern Taiwan. Tectonophysics 125:129–146

Ma KF, Lee CT, Tsai YB, Shin TC, Mori J (1999) The Chi-chi, Taiwan earthquake: large surface displacements on an island thrust fault. EOS Trans 80:605–611

Seno T, Stein S, Gripp AE (1993) A mode for the motion of the Philippine Sea plate consistent with NUVEL-1 and geological data. J Geolphys Res 98:17941–17948

Soil Survey Staff (1994) Keys to soil taxonomy, 6th edn. U.S. Dep. Agric., Soil Conservation Service, Washington, DC

Suppe J (1981) Mechanics of mountain building in Taiwan. Mem Geol Soc China 4:67–89

Teng LS (1990) Geotectonic evolution of late cenozoic arc-continent collision in Taiwan. Tectonophysics 183:57–76

Teng LS (1996) Extensional collapse of the northern Taiwan mountain belt. Geol 24:949–952

Tsai YB (1986) Seismotectonics of Taiwan. Tectonophysics 125:17–38

Tseng MH, Liew PM, Chi WR, Shih TS (1992) Pollen analysis of the Tananwan formation, northern Taiwan. J Geol Soc China 35:247–259

Wang ST, Wang MK, Sheh CS (1988) The 1988 soil map of Taiwan. In: Proceedings of the fifth international soil management workshop: classification and management of rice-growing soils, 11–23 Dec 1988, Taichung, Taiwan

Yu SB, Chen HY, Kuo LC (1997) Velocity field of GPS stations in the Taiwan area. Tectonophysics 274:41–59

2.1 History

The soil survey of Taiwan has been developed for more than 90 years. The soil survey program was started in Taiwan very early; Dr. Shibuya published the reconnaissance soil texture map in 1911 (Fig. 2.1) and the soil reaction map in 1913 (Fig. 2.2). After the reconnaissance soil survey, a series of county soil survey programs were conducted until World War II. The reports focused on soil fertility, land use, and rural infrastructure planning.

After World War II, two major programs on soil survey were conducted, i.e., "Key to the soils of Taiwan" and "A General Study on the Soil Fertility of Taiwan" (Si and Chang 1951) and "County's soil survey" (1947–1964). The soil classification system followed the 1938 USDA yearbook system of soil classification. Some special purpose soil survey programs were conducted during this period, such as marginal lands land use capability survey (1960) and soil fertility survey of farmlands (1967, 2008), fertility capability classification (1983) for land use planning, and fertilizer management. The aims of most of the programs were concentrated on soil fertility topics due to lack of fertilizers and food security consideration during the period (Guo et al. 2002).

The soil information system can provide information about the pattern of the soil cover and its characteristics for us to analyze and display the topics on soil resources management. Soil information system of plain region and hill lands was constructed by the Taiwan Agricultural Research Institute (TARI) and Soil and Water Conservation Bureau (SWCB) in 1988 and 1990, respectively. The soil information application has become very popular since then. The soil data provided for application in the university and by the government was applied in the fields of environment protection, land management, nature resources management, and civil engineering application. More than 30 applicants from different institutes ask for soil information every year.

Soil survey has been conducted since the 1920s; some Japanese soil scientists made an approximate soil survey including soil pH and soil texture maps. Below are 19 survey projects that can be listed as important events in the last 50 years since the 1950s.

- Reconnaissance soil survey of Taiwan (1946–1952), conducted by TARI, Council of Agriculture.
- Soil survey and soil fertility evaluation, conducted by Taiwan Fertilizer Company (TFC).
- Soil Survey of Taoyuan County, conducted by TFC.
- Soil survey of Tobacco production of Taiwan, conducted by Department of Agricultural Chemistry, College of Taiwan Provincial Agriculture.
- Soil survey of agricultural farms of Taiwan Sugar Company (TSC), conducted by TSC.
- Soil survey for land use of agriculture and forestry, conducted by TARI.
- Soil survey of Tea production in Kuang-See Township, conducted by the Department of Agricultural Chemistry, National Taiwan University (DAC/NTU).
- Detail survey of Taiwan saline soils, conducted by Department of Agricultural Chemistry, College of Taiwan Provincial Agriculture.
- Soil survey and land use of Lu-Kang region of Changhwa County, conducted by TFC.
- Soil survey of saline soil along seashore, conducted by TSC.
- Detailed soil survey of Taiwan rural soils, conducted by National Chung Hsing University (NCHU) from 1962 to 1976 and by TARI from 1974 to 1979.

Z.-S. Chen et al., *The Soils of Taiwan*, World Soils Book Series,
DOI 10.1007/978-94-017-9726-9_2, © Springer Science+Business Media Dordrecht 2015

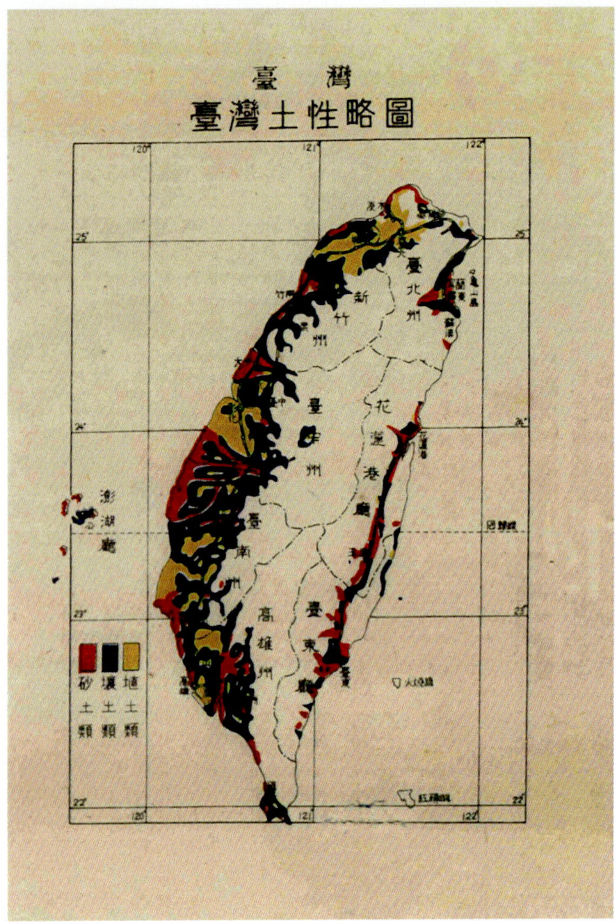

Fig. 2.1 Soil texture map of Taiwan by Dr. Shibuya in 1911. Copyright by Prof. Chen

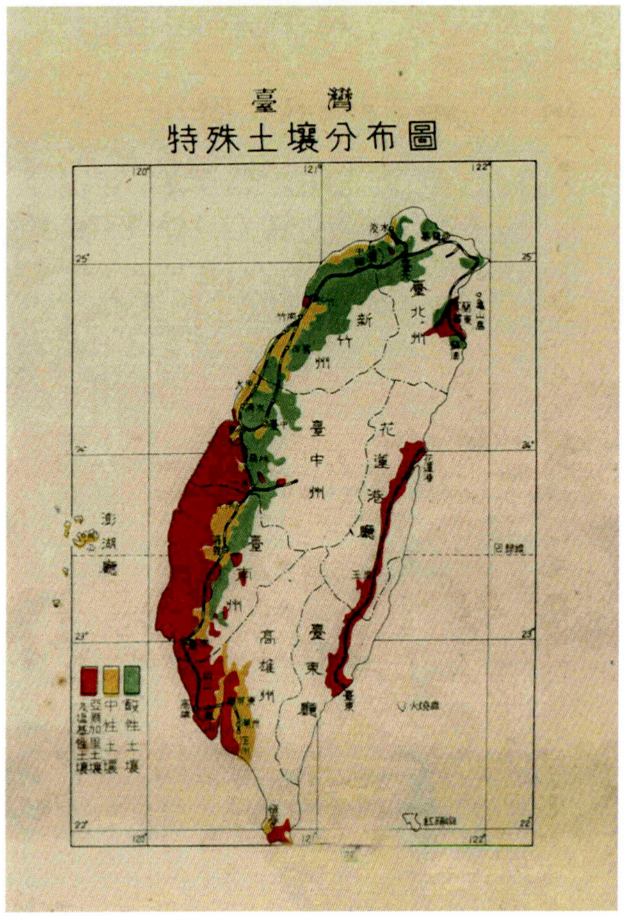

Fig. 2.2 Soil reaction map of Taiwan by Dr. Shibuya in 1913. Copyright by Prof. Chen

- Soil survey of Southern region of Taipei Basin, conducted by TFC.
- Reconnaissance soil survey of Taiwan Forest soil, conducted by Forestry Bureau of Taiwan Province.
- Detailed soil survey of Taiwan hill land (1980–1988), conducted by Bureau of Agriculture and Livestock of Hill Land, Taiwan Province.
- Survey of contaminated soils (1984–2000), conducted by NCHU, NTU, Taiwan Agricultural Chemical and Toxic Materials Research Institute (TACTRI), and National Pingtung University of Science and Technology (NPUST).
- Survey of soil productivity grade of upland in Taiwan (1986–1990), conducted by Department of Soil Science, NCHU.
- Detailed soil survey of Taiwan forest soils, which the elevation ranged from 1,000 to 4,000 m (1993–2002), conducted by Taiwan Forestry Research Institute (TFRI), Council of Agriculture.
- Detailed grid soil survey of Taiwan rural soils (2000–2010), conducted by TARI.

- Soil quality monitoring project after different soil management system (1997–2007), conducted by TARI.

2.2 Survey Report and Mapping

Systematic soil surveys of agricultural lands, hill lands, and forest lands on a detailed scale (1:25,000) were implemented in Taiwan in the period 1960–2006. The programs were conducted by NCHU and TARI for farmlands from 1960 to 1978, SWCB was in charge of the hill lands region during 1979–1988, and Taiwan Forestry Research Institute (TFRI) and Bureau of Forestry (BOF) responded for forest region from 1994 to 2006. The soil characteristics of Taiwan were well understood after a 46-year period of soil survey projects. However, different institutes conducted these three stages of soil survey programs, so it takes a lot of time to correlate the soil attributes, map units, and boundaries of soil delineations between these different programs. Unclear soil mapping concepts and techniques in those days also makes soil correlation a problem (Guo et al. 2002).

Eight soil survey projects by 19 teams abovementioned have made valuable and meaningful contribution leading to improvement of human life, such as in agricultural development, designing agricultural policy on agriculture and environment, and education. Especially, the great contribution of soil survey on Taiwan agricultural development is realized as a knowledge infrastructure.

2.2.1 Reconnaissance Soil Survey of Taiwan Province (1946–1952)

This survey was conducted by TARI from 1946 to 1952. Eight volumes of soil survey reports, including eight counties of Taichung, Penghu, Tainan, Kaohsiung, Pingtung, Taoyuan, Hsingchu, and Miaoli, and soil maps on the scale of 1:100,000 have been published. Unfortunately, four counties of soil survey reports were not published owing to limitations in budget. This soil survey followed the soil taxonomy created by USDA in 1938, and soil mapping unit was soil series, soil types, and soil phase for different soils in this stage. The soil survey quality was regarded as a very high soil survey work at that time.

2.2.2 Detailed Soil Survey of Taiwan Rural Soils (1962–1976, 1974–1979)

This survey was conducted over a period of 15 years by the Department of Soil Science, NCHU from 1962 to 1976, with a team of 12 soil surveyors. It was chaired by Professor Min-Kao Wang. Seven volumes of soil survey reports were published for the area of southern Taiwan, including seven counties of Changhwa, Tainan, Pingtung, Chayee, Kaohisung, Yulin, Taichung, and Nantou, and 90 soil maps on the scale of 1:25,000 were published. There are 342 soil series established by the team of soil survey in this stage.

Then this soil survey was taken over by TARI for 6 years from 1974 to 1979, chaired by Researcher Chun-Chuang Chen. About 10 soil surveyors joined this team. Four volumes of soil survey reports, including four counties of Taoyuan, Miaoli, Hsingchu, Taipei, and Ilan, and 88 soil maps on the scale of 1:25,000 were published. There are 278 soil series established by the team of soil survey in this stage.

The most important contribution of this detailed soil survey is to establish 620 basic soil series of Taiwan rural soils and to publish 177 soil maps on a scale of 1:25,000 for soil management and fertilization recommendation. The soil series was identified by the differences in parent materials, drainage classes, texture changes through the profile, and soil reaction in some cases. The soil survey system was followed by the Soil Taxonomy system created by USDA in 1949, and soil mapping unit is the soil type for rural soils. This soil survey work was regarded as the most important contribution for agricultural production and development in Taiwan.

2.2.3 Detailed Soil Survey of Taiwan Hill Land (1980–1988)

This survey was conducted by the Bureau of Agriculture and livestock of hill land (reorganized as SWCB later), Taiwan, from 1980 to 1988. This soil survey was chaired by Researcher Chun-Chuang Chen, who was working at TARI. About 12 soil surveyors joined this team. The survey area of hill land is the elevation lower than 1,000 m except the rural soils surveyed before 1980. A total of 13 volumes of soil survey reports of Taiwan hill lands and 215 soil maps on the scale of 1:25,000 were published. There are 432 soil series established by the team of soil survey in this stage, in about 10 years (1980–1988).

The soil series was also identified by the differences in parent materials, drainage classes, texture changes of the profile, and soil reaction in some cases. The soil survey system was followed by the approximate Soil Taxonomy created by USDA revised in 1960, and the soil mapping unit is soil phase including surface soil textures and slope. This soil survey work was regarded as the most important contribution for Taiwan agricultural production and development.

2.2.4 Soil Survey of Productivity Grade of Taiwan Upland (1986–1990)

This survey was conducted for 5 years by the Department of Soil Science, NCHU from 1986 to 1990. This soil survey was chaired by Professor Chur-Chung Yang, who was working at the Department of Soil Science, NCHU. About 10 students joined this team. Four classes of soil productivities were established depending on the existence of texture changes of soil profile and gray mottles formed at different soil depths, such as 40–60, 60–90, 90–150 cm, and lower than 150 cm from soil surface. Finally, 168 soil maps of soil productivity classes of Taiwan upland on a scale of 1:25,000 were published. The conclusion is that the worst soil productivity corresponds to the occurrence of gray mottle which was produced in the surface 40–60 cm depth, while the best soil productivity is the gray mottle in soil deeper than 150 cm.

2.2.5 Soil Survey of Contaminated Soils (1984–2000)

This survey of contaminated soils in Taiwan was conducted for 10 years by NCHU, NTU, Taiwan Agricultural Chemical and Toxic Materials Research Institute (TACTRI), and National Pingtung University of Science and Technology (NPUST). This soil survey was chaired by Professor Yin-Po Wang, who worked at the Department of Soil Science, NCHU. About two researchers from government agencies, five professors of universities, and 10 students and research assistants joined this team. Owing to the serious soil contamination found around the discharged area of 90 industrial parks in Taiwan, Taiwan Environmental Protection Administration (Taiwan/EPA) started the soil survey of contaminated rural soil by heavy metals from 1984 to now, including As, Cd, Cr, Cu, Hg, Ni, Pb, and Zn.

More than 10,000 soil sampling locations, 150 potential polluted sites, and five seriously polluted sites in Taiwan were included into this GIS system, hence we can make an evaluation of the status of contaminated sites to identify the relationships between the contamination soil status and contaminated resources, and also to request the best remediation techniques to be applied in the remediation projects in the future.

2.2.6 Detailed Soil Survey of Taiwan Forest Soils in Mountains (1993–2002)

This detailed forest soil survey was conducted for 10 years by Taiwan Forestry Research Institute (TFRI) from 1993 to 2002. This soil survey was chaired by Mr. Kuang-Chin Lin, who was working at the Department of Silviculture, Taiwan Forest Research Institute, Council of Agriculture (TFRI). About 16 soil surveyors joined this team for 10 years. Until now, more than 10 volumes of soil survey reports, and more than 200 soil maps on a scale of 1:50,000 were published. More reports and soil maps will be continuously published in the next few years. The soil data and soil maps can be retrieved from the Internet (http://ims.tfri.gov.tw/tfri1/main. aspx/) (TFRI 2005).

The soil classification of this survey was based on *Keys to Soil Taxonomy* (Soil Survey Staff 1998), and soil mapping unit is soil phase including surface soil textures and slope. This soil survey work was regarded as the most important contribution for Taiwan forestry production and development in the next decade. All the published reports and soil maps were combined into the Taiwan National Soil Information System in the website of TARI (http://www.tari.gov. tw/index_in.htm) (Guo et al. 2005), shown in Chinese, not included English version.

2.2.7 Detailed Grid Soil Survey of Taiwan Rural Soils (2000–2010)

Taiwan Agricultural Research Institute (TARI) began to conduct a detailed grid soil survey project from 2000 to 2010. This soil survey was chaired by Mr. Horng-Yu Guo (HYGuo@wufen.tari.gov.tw), who is working at the Department of Agricultural Chemistry, Taiwan Agricultural Research Institute, Council of Agriculture (TARI/COA) (Guo et al. 2000). One representative soil profile was sampled in every 250 m based on the air photo pictures on a scale of 1:5,000. This means there is one representative soil profile data that was sampled in 6.25 ha by soil auger from soil surface to 150 cm depth, including six samples in one profile at 0–15, 15–30, 30–60, 60–90, 90–120, and 120–150 cm depth from soil surface. The soil samples showed the soil morphological characteristics, and also the basic physical, chemical, and biological properties in the laboratory of TARI.

More than 5,000 soil profiles (or 30,000 soil samples) were sampled in one county, and more than 90,000 soil profiles (or 540,000 soil samples) database in the whole of Taiwan were included in the GIS system of National Soil Information System of Taiwan. From this valuable database, many item maps were produced and published by TARI for soil interpretation, land planning, and soil management, especially for fertilizer recommendation for different crops production (http://www.tari.gov.tw/index_in.htm). Until now, this website is shown in Chinese, not in English version. These maps are also demonstrated in many workshops of reasonable fertilization workshop on high economic fruits in Taiwan, especially for rice, corn, sorghum, citrus, wax apples, mango, watermelon, apple, and vegetables. The soil item maps are listed as follows:

- Maps of soil pH values of surface soils
- Maps of soil texture classes of surface soils
- Maps of soil organic matter content of surface soil
- Maps of soil drainage classes
- Maps of soil erosion prediction
- Maps of crop suitability classes for 300 crops and economic fruits
- Maps of soil moisture regimes
- Maps of soil temperature regimes
- Maps of Soil Orders based on Soil Taxonomy (Soil Survey Staff 1999)
- Maps of soil available nutrients including N, P, K, Na, Ca, Mg, Fe, Mn, Cu, and Zn

- Maps of soil management groups
- Maps of bioavailable heavy metals (extracted by 0.1 M HCl).

2.2.8 Soil Quality Monitoring Project After Different Soil Management System (1997–2007)

The Taiwan Agricultural Research Institute (TARI) conducted a soil quality monitoring project from 1998 to 2008. This soil quality monitoring project was chaired by Mr. Horng-Yu Guo (HYGuo@wufen.tari.gov.tw), working at the Department of Agricultural Chemistry, Taiwan Agricultural Research Institute, Council of Agriculture (TARI/COA) (Guo et al. 2003). One hundred and fifty rural soil sites were selected from different environmental conditions including different soils, crops, and climates conditions, for monitoring the changes in soil quality in each site. The area of each representative site is 1 ha. Thirty soil samples were sampled in this 1 ha site and soil characteristics are also analyzed for evaluation of changes in soil quality.

This monitoring project was conducted in the second round of 150 sites in Taiwan (first round in 1997–2002 and second round in 2002–2007). They have selected some soil indicators to evaluate the change in soil quality, including soil texture, pH, organic C, total N, exchangeable K, Na, Ca, Mg, and Mn (extracted by 0.1 M sodium acetate at pH 7), soluble Fe, Mn, Cu, Zn, Cd, Pb, Cr, and Ni (extracted by 0.1 M HCl), and available P, K, Ca, Mg, Na, B, Cu, Zn, Fe, and Mn (extracted by Mehlich #3 solution). Unfortunately, no data of biological soil indicators were included into their database. After 5 years' monitoring, the results indicate that the soil-available phosphorus content is much higher than the need for crops and soil-available K is still not enough for crop production. We need more workshops to train farmers to have better soil and fertilizer management in the future.

2.3 Soil Information System (Application and Education)

Soil information was applied on several fields by TARI, such as: the soil productivity classification of paddy fields (1986), soil productivity promotion planning (1993), soil management grouping (1998), assessment of susceptibility of soil liquefaction (1999), assessment of the susceptibility of nitrogen fertilizer (2000), and soil quality monitoring (2001). An agricultural environmental management expert system (rice) is in progress; the system includes climate data, soil data, and soil testing data, cadastral data, and crop management of knowledge base provides recommendations for cultivation process for each farmer.

There are 15 attributes data for each map unit in this system that include: parent material, soil morphology (color or special feature), soil formation, soil drainage class, slope class, soil reaction, soil calcareous properties, soil type, four layers of different depth of soil texture class of the representative profile, and soil phases (salinity, stoniness and gravelly). The attributes of soil units except soil reaction are qualitative, i.e., by grades or classes classification. TARI has applied these soil databases of Taiwan farmlands to recognize the distribution and to assess the grade of low-productivity croplands.

However, the basic soil attribute data that soil survey reports provided could not be satisfied with the requirement of intensified farming system operation and rapid-changing soil quality (Guo et al. 1992). A soil fertility project, which concerns plant nutrients and heavy metals contained in soil is in progress. These attribute data will be as compensation for the soil survey data shortage of the management of soil nutrients and food safety protection. This chapter provides experiences of providing information for the management of low-productivity and degraded soils that we have done in Taiwan.

2.3.1 Soil Information Apply to Low Productivity Land Assessment

By the concept of FAO land qualities, the low-productivity soils assessment chose several land qualities that criteria could link with the soil information system of Taiwan farmland, such as shallowness, stoniness, workability, salinity, acidity, water logging, droughty (moisture availability), nutrients deficiency. The low-productivity soils criteria inferred from the soil survey reports, research reports, and derived soil quality and fields experience. The soil database was evoked in relation with criteria to assess and map the distribution of the low-productivity soils.

The soil properties and soil fertility in Taiwan are not of good quality; however, many crops have good yields; intensive fertilizer application and soil management are the reasons for the results. Though improved management and ameliorative measures have been intensively studied to solve these low productivity soils, the problems of economical feasibility and social requirement prohibit the program of soil reclamation proceeding.

Soil information and soil management system provide a clear understanding of low productivity soils and degradation soils in Taiwan farmlands (Table 2.1). It provides an overview of the relative extent of physical resource limitations to agriculture and other forms of land use in the whole

Table 2.1 The soil attributes of soil information applied on the assessment of low productivity soils in Taiwan

Low-productivity soils	Attributes used for assessment	Acreages (ha)
Shallowness	Soil types; profile depth; soil phase	137,610
Stoniness	Soil phase	28,390
Tillage practice and soil workability	Soil type; profile texture; drainage class	107,010
Salinity	Soil salinity; drainage class	62,970
Acidity	Soil formation and pH	190,830
Water logging	Soil type; drainage class	47,480
Droughty	Profile texture; drainage class	55,970
Nutrients deficiency	Parent materials; pH; calcareous materials	268,980

Copyright by Prof. Chen

country; highlight the area which calls for the treatment or management of specific land resources constraints, so the regional or national action plans can be better focused on specific problems; indicates the limitations of the data and the amelioration methods, and hence the priority needs for improved information and research.

2.3.2 Soil Information Apply to Soil Liquefaction Assessment

Soil liquefaction has been observed in many places after the 921 Taiwan Chichi Earthquake mentioned in Chap. 1. Soil liquefaction resulted in disastrous effects such as tilted buildings, foundation failure, and sand volcanoes or sand boils in the fields. Because liquefaction only occurs in saturated soil, its effects are most commonly observed in low-lying areas near water bodies such as rivers, lakes, bays, and oceans. Parts of Taichung, Nantou, Changhua and Yunlin's Plain regions belong to these kinds of landscapes. The Tzo-Sui Alluvial Fan covers the most area in this region. The Tzo-Sui River had changed its waterway several times in the past 200 years, so many abandoned channels spread around this region. Taichung, Changhua, and Yunlin plains are also near the seashores, these lands have high potential of soil liquefaction.

Sands were considered to be the only type of soil susceptible to liquefaction, but liquefaction has also been observed in gravel and silt. So different soil textures are of different sensitivity to soil liquefaction. Soil drainage classes can indicate the different levels of soil water contents, which can be also used to evaluate the sensitivity of soil liquefaction. We evaluated, classified, and mapped the susceptibility of soil liquefaction in disastrous regions based on three categories in soil information. The map further verified the

reliability in earthquake areas. Most of the boundaries of the soil liquefaction susceptibility units match the soil liquefaction regions. The reliability of the assessment of soil liquefaction susceptibility is very high. The information provided here could be very helpful in the developing plan for evading soil liquefaction hazard.

2.3.3 Soil Information Apply to Food Safety

The quality of food (rice) has become a social-economic issue of concern in Taiwan due to soil pollution. New insight into the relationship between soil quality and the quality of food should be used by soil environmental scientists around the world to improve scientific methods to provide tools to estimate risks, set soil standards, and reduce the uptake of heavy metals by rice. To assess the risk of the metals in these soils, soil quality guidelines standards have been developed in many countries, including Taiwan. Such soil quality standards are based on the total heavy metal content although research results from the last decade clearly indicate that the risk of heavy metals in soils does not depend on the total metal content only. In fact, a part of the total metal content can be considered nonreactive.

At 16 sites, 12 different cultivars of rice were used, including the ones most commonly grown at present in Taiwan. This includes japonica type, *indica* type, and sticky rice which are the three major rice varieties. It was found that the heavy metal uptake rate differs among the rice varieties: uptake by *indica* varieties is higher compared to *japonica* varieties, which is especially true for cadmium in brown rice. It was commonly observed that although the soil cadmium concentration was below the standard level of monitoring, the cadmium concentration in the brown rice of several *indica* varieties exceeded the rice safety regulation for cadmium. A detailed report can be referred from Guo et al. (2007). The regression equations to predict cadmium uptake by brown rice from soil properties according to equation are calculated. The results indicate that the prediction of brown rice of cadmium concentration is by considering soil pH, organic carbon, and CEC.

Experimentally derived models were used to calculate the so-called risk map for rice cropping. This was done by calculating the critical levels of cadmium in the soil above which the rice cadmium content exceeds the food quality standard. The soil database from Taiwan was used to compare this critical level with the actual measured values across the country. The results indicate that areas exist where the quality of *indica* species will be insufficient (i.e., cadmium in the rice will exceed the standard of 0.2 or 0.4 mg kg^{-1}). Such areas include soils derived from marine sediments as well as contaminated soils. One of the options that are rather easy to implement by farmers is to grow *japonica* genotype rice in these regions.

2.3.4   Soil Museum for Soil Function Education

The international level of national soil museums were established and opened in 2005 at TARI for service to students, farmers, teachers, and government officials (http://www.tari.gov.tw/index_in.htm) (Fig. 2.3). More than 200 soil monoliths were displayed in this museum (Fig. 2.4).

The small scale of university level soil museum was established and opened in 2005 at National Taiwan University (NTU Soil Museum) (http://Lab.ac.ntu.edu.tw/soilsc/) (Figs. 2.5, 2.6 and 2.7), which provided more than 100 soil monoliths to show the typical soil morphological characteristics and valuable posters to display the soil properties and to use these characteristics for more interpretation in soil

Fig. 2.3 Opening ceremony of Taiwan Soil Museum in 2005. Copyright by Prof. Chen

Fig. 2.4 The display of soil monoliths in the Taiwan Soil Museum. Copyright by Prof. Chen

Fig. 2.5 The display of soil monoliths in NTU Soil Museum. Copyright by Prof. Chen

Fig. 2.6 Soil education for new generations in Taiwan (the standing lecturer is the primary author of this book in the Taiwan Soil Museum). Copyright by Prof. Chen

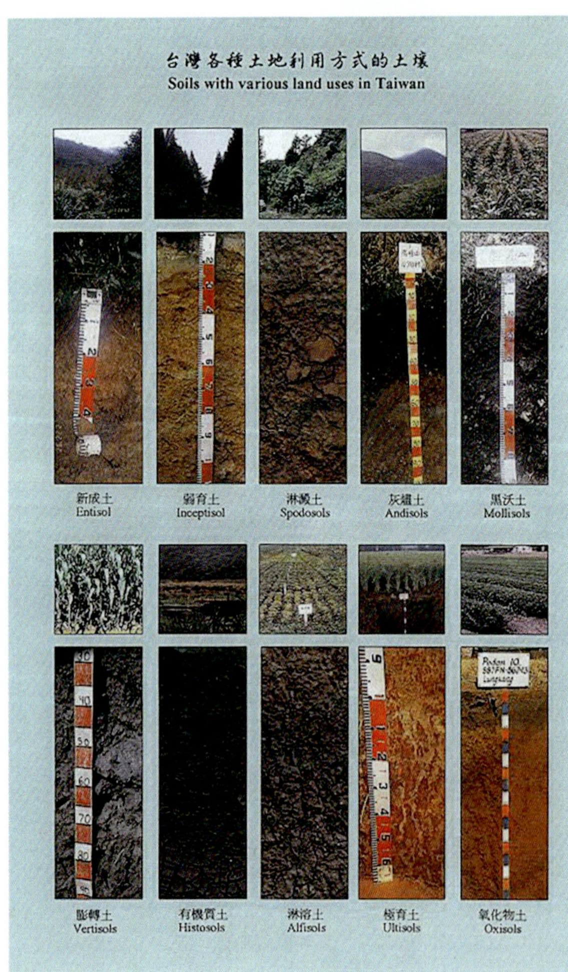

Fig. 2.7 Ten major soil orders of Taiwan soils are one of the typical posters shown in the Soil Museum of National Taiwan University. Copyright by Prof. Chen

management and soil quality, crop quality, and environmental quality in the future (Chen 1998; Chen et al. 1999a, b; Lee et al. 2006).

2.4 Pedology Studies in Taiwan

2.4.1 The Soil Characteristics and Genesis of Soils Derived from Different Parent Materials (1960–1975)

The basic research in soil morphological characteristics, clay mineralogy, and pedogenic processes of different representative soils derived from parent materials were conducted by Prof. Chang working at National Taiwan University (Chang 1962, 1964, 1966, 1969, 1971, 1972, 1974, 1975). The basic database of the representative "major Soil Groups" of Taiwan provided valuable soil information during the 1960s and made great contribution toward the agricultural development of Taiwan. These soil groups included sandstone alluvial soils, slate alluvial soils, schist alluvial soils, brown forest soils, Podzolic soils, volcanic ash soils, and red soils in the terrace of Taiwan.

2.4.2 The Soil Characteristics and Genesis of Rice-Growing Soils of Ultisols (1980–2005)

This series of studies was conducted for 10 years by colleagues from National Taiwan University from 1980 to 2005, especially for Ultisols with plinthites. Till date, more than 10 scientific papers have been discussed and published in international journals, such as Soil Science Society of America Journal and Soil Science, and related local journals in Taiwan (Chang and Chen 1982; Chang et al. 1983; Chen 1984; Chen and Chang 1984, 1985; Chiang et al. 1997; Hseu and Chen 1994, 1995, 1996, 1997, 1999, 2000, 2001; Hseu et al. 2000; Jien et al. 2004).

2.4.3 The Soil Characteristics and Genesis of Podzolic Soils (1988–2005)

These series studies of Podzolic soils in Taiwan are conducted since the 1960s. There are significant albic horizons formed in subalpine and alpine forest soils in the central ridge of Taiwan, including Mountains of Alishan, Taipingshan, Chusahn, where the elevation is always higher than 2,500 m above sea level, with high precipitation (>4,500 mm/year and low annual air temperature, <15 °C). The second stage of these series studies was conducted by colleagues from the National Taiwan University from 1988 till the present. Four typical Podzolic soils in Taiwan were proposed (Chen and Tsai 2000), including Spodosols, Ultisols, and Inceptisols with or without placic horizon based on the Soil Taxonomy (Soil Survey Staff 1999, 2003; Hseu et al. 1999b).

Until now, more than 10 scientific papers have been discussed and published in international journals such as Soil Science Society of America Journal, Soil Science, Geoderma, and related local journals of Taiwan (Chen et al. 1989, 1995; Chen and Tsai 2000; Hseu et al. 1999, 2004; Li et al. 1998a, b; Lin et al. 1988, 2002; Liu and Chen 1990, 1991, 1998, 2004; Liu et al. 1998; Wu and Chen 2005). The most significant soil characteristics of these Spodosols, Ultisols, or Inceptisols, are loamy soil texture in the Spodic horizon or Spodic morphological horizon, whose clay content ranged from 250 to 350 g/kg. The other characteristics are that the soils have high precipitation, especially it is higher than 4,500 mm/year in the summer. These special characteristics are different compared with other countries which have the distribution of Spodosols.

2.4.4 The Soil Characteristics and Genesis of Volcanic Soils (1990–2005)

The first stage study of volcanic ash soils in Taiwan was conducted in the 1970s by Prof. Chang working at National Taiwan University (Chang 1972, 1974, 1975). Some basic research can be compared with the reports from Japan volcanic soils (Saigusa et al. 1991; Shoji et al. 1982, 1993). After 15 years, the second stage study was conducted for another 15 years by colleagues from National Taiwan University from the 1990s till the present, especially for non-

allophanic soils. Until now, more than 15 scientific papers have been discussed and published in international journals such as Soil Science and related local journals of Taiwan (Asio and Chen 1998; Asio et al. 2000; Chen and Huang 1991; Chen et al. 2001; Huang and Chen 1990, 1992; Huang et al. 1988, 1993, 1994).

Until now, 86 soil pedons have been studied in whole National park, major distributed in Mountains of Samao, Chishing, and Tatung. The soils were collected from different transects including Andisols (Hapludand, Fulvudand or in terms of Allophanic Andisols) and Inceptisols (Dystrudept, or in terms of nonallophanic Andisols) are formed in different regions of the park. The Inceptisols are distributed in the lower elevation and sharp landscape of the study area, about 60 % of the total area and the Andisols are distributed in the higher elevation and stable landscape of the study area, about only 15 % of the total area. The relationship between volcanic soil genesis and environmental factors is difficult to recognize in this area. We estimated that the interaction within climate, landscape, and vegetation would be variable with time and reduced the effect of landscape on the volcanic soil genesis.

Soil pedons selecting from Mt. Tatung to lower terrace were collected to understand the pedogenic process of volcanic soils and to identify the soil indicators of transition soils from Andisols to Ultisols under the same parent materials. The results of this study suggest that six pedogenesis indices may be useful to identify the transitional soils among Andisols, Inceptisols, and Ultisols, including soil texture, soil consistence, bulk density, phosphate retention, and amorphous material content of Alo + 1/2Feo (%), and Sio (%) (Tsai and Chen 2006; Tsai et al. 2006).

Volcanic activities occurred from about 0.5 to 0.8 million years ago, leaving various signs of post-volcanic activity that we can see today. The Pleistocene volcanic activities are most important in the geologic evolution of Taiwan. The most typical characteristics of Taiwan volcanic soils were collected from different transects including Andisols (Hapludand and Fulvudand or in terms of Allophanic Andisols) and Inceptisols (major with Dystrudept or in terms of nonallophanic Andisols) are formed in different regions of the park.

2.4.5 The Soil Genesis and Geomorphology on River and Marine Terraces (2005–Present)

Most soil properties are time-dependent variables, which can be used as indicators of the duration of pedogenesis. The island of Taiwan has experienced mountain building since about 5 Ma, and is still active until the present time (Ho 1988). River terrace is one of the most prominent geomorphic features on the land surface serving as geomorphic marker to gauge the differential or absolute surface deformation. However, such use of river terrace is often hampered by the absence of well-documented ages for deformed or partially preserved surfaces (Tsai and Sung 2003).

The Pakua tableland defines a deformation front of the fold-and-thrust belt in central Taiwan due to tectonic compression (Lee et al. 1996; Delcaillau et al. 1998). There is a series of widely unpaired river terraces developed in the south of the tableland. They are divided into as many as six altitudinal levels. A soil chronosequence consists of genetically-related suites of soils of varying age that evolved under similar conditions of vegetation, topography, and climate (Harden 1982). The surface deposits of these river terraces may form a soil chronosequence since they have comparable soil-forming factors except the age of the deposit. However, the identification of a particular set of soil properties that are consistently indicative of soil development is needed. Therefore, Tsai et al. (2006) and Tsai et al. (2007a) used six representative soil pedons from the six levels of terraces on the tableland (from the highest pedon PK-1 to lowest pedon PK-6) to characterize the soil properties in a chronosequence, and to relate the pedogenic processes in the major terraces to the formation and evolution of the landscape. The soil morphological, physical, and chemical properties as well as the clay mineral variation showed that pedogenic intensity is strongly dependent on the terrace levels with varying formation age. Based on the crystallinity ratios of free iron, the soils give an estimated age of 40–400 ka for the river terraces of the tableland. The soils can be divided into three domains as Kandiudult for pedon PK-1, Paleudult (or Hapludult) for pedons PK-2, -3, -4, and -5 and Dystrudept for pedon PK-6, based on *Soil Taxonomy*. The degree of soil development increases with altitude in a sequence from PK-1 to PK-6 forming a post-incisive type of soil chronosequence in accord with the evolution of the geomorphic surface by successive river incision in the study area.

Tsai et al. (2007b) further explored the lateritic river terraces on the Pakua, Chushan, and Touliu tablelands in central Taiwan whose geomorphic features are inadequate for regional terrace correlation. They used an alternative pedogenic scheme based on the weighted mean profile development index (WPDI). Four soil pedons from the Chushan and Touliu terraces are classified as Inceptsol and Ultisol in *Soil Taxonomy*. By citing another six soil pedons (PK-1 to -6) of the Pakua terraces in the previous study (Tsai et al. 2006), the WPDI values of all the total 10 soil pedons suggest a chronological order for the terrace correlation. The correlation suggests the Chushan and Touliu terraces likely belong to one anticline as a whole, and is separated by the Chinshui River. This anticline represents the southern extension of the Pakua anticline resulted from the frontal thrusting of the Changhua Fault.

Placing a time or age constraint on soil development has become one of the most important factors in the disciplines of soil science and fluvial geomorphology. The radiocarbon (^{14}C) method is unable to date old terraces or soils, but the longer half-life of cosmogenic beryllium-10 (^{10}Be) permits this isotope to be useful to determine soil ages and rates of formation. Red soils (redder than 7.5YR) are commonly distributed on old, high altitude river terraces throughout Taiwan. Tsai et al. (2008) used three pedons of alluvium-related red soils were sampled from the Taoyang (TY-YM), Pakua (PK-1), and Chiayi terraces (CY-1), respectively located in northern, central, and southern Taiwan. Meteoric ^{10}Be dating shows ages of $\geq$261 ka for TY-YM, $\geq$124 ka for PK-1, and $\geq$386 ka for CY-1. However, the trend of these ages does not agree with the degree of soil development. The former two ages are likely underestimated through loss of ^{10}Be due to strong leaching and considerable erosion. The age obtained from CY-1 may represent the minimum time required for soils develop into Oxisols in Taiwan.

The Tadu tableland is located at the deformation front of the Western Foothills in central Taiwan (Fig. 1.3). Previous geomorphic studies suggested a fluvial terrace landform developed on the surfaces of the tableland. However, other suggestions have been proposed in recent years. Tsai et al. (2010) solved the argument by examining the pedogenesis of surface deposits. Five soil pedons were sampled to the depths of C horizons in the surface deposits of the tableland. These soils were classified as Paleudult and Kandiudult of Ultisols according to *Soil Taxonomy*. In addition, their morphological characteristics were further quantified in terms of soil development by the horizon index and WPDI. The WPDI values of 0.64–0.73 for the soils of the Tadu tableland agree with those of Ultisols developed on the Pakua tableland. Both qualification and quantification assessments indicate all soils of the Tadu tableland are equivalent in terms of degree of pedogenesis. As a result, all the geomorphic surfaces on the tableland probably developed at the same time, which rules out the interpretation of fluvial terraces developing on the tableland. The pedogenic evidence supports the recently proposed kinematic model for the development of the Tadu tableland. Moreover, relative dating based on regional pedogenic correlation suggests the folding activities of the Tadu tableland occurred more recently than those of the Pakua tableland did.

Vertisols, Mollisols, and Entisols are generally found on different levels of marine terrace herein, but no detailed investigations in soil chronosequence have been conducted by integrating field morphology, physiochemical characterization, micromorphology, and mass balance interpretations.

Tsai et al. (2007c) found that strongly developed angular blocky structures, pressure faces, and slickensides are more common in higher terrace soils than in lower terrace soils. In their studies, including depth to C horizon, solum thickness, and thickness of the clay-enriched zone show increase with relative terrace age. Although only one to two profiles per terrace were characterized, the following soil analytical characterizations increase with time: the degree of sand grains weathering, pH (H_2O), organic carbon, CEC, contents of Fe_d, Fe_o and Mn_d. Soils on all terraces have a mixed mineralogy. Mica, smectite, and kaolinite have slightly increased with increasing terrace age. Furthermore, the dominant processes identified with mass-balance analysis include loss of bases (Ca and Mg), iron, and clay with time. The soil properties, including analytical and mineralogical characterizations, which do not have notable changes with time are primarily due to relatively young soil age (<20 ka). Moreover, Huang et al. (2010) indicated that the relatively weaker degree of soil development on the oldest terrace at the southern part of Hua-tung marine terrace supports the conclusion that uplift rates in the southern area are greater than at middle and northern parts of Hua-tung marine terraces in eastern Taiwan (Fig. 7.2).

2.5 Work Needs in the Future

Soil mapping quality should be improved for more and more public requirements. The present soil database preferred chemical properties, but important physical properties like bulk density and hydraulic conductivity need to be collected for further application. The basic soil attributes data in soil survey reports could not be satisfied with the requirement of intensified farming system operation and rapid-changing soil quality. Therefore, we need to develop new strategies of soil survey for the future. The major items will be discussed, including the development of new soil survey techniques to identify the boundary of contaminated soils, improvement in soil properties due to the land use changes from paddy soils, use of national soil information system for more soil interpretation, and continuous education in soil function.

References

Asio VB, Chen ZS (1998) Micromorphological study of andesite weathering under the perhumid subtropical climate of north Taiwan. Taiwan J Sci 13:259–269

Asio VB, Chen ZS, Tsou TC (2000) Comparison of clay dispersion pretreatments and the estimation of clay content in volcanic soils of Taiwan and the Philippines. Food Sci Agric Chem 2:20–27

Chang CM (1964) Profile and clay mineral studies of Taiwan alluvial soils. J Chin Agric Chem Soc 2:109–122 (in Chinese, with English summary, Tables and Figures)

Chang CM (1966) Profile and clay mineral studies of older alluvial soils in Taiwan. J Chin Agric Chem Soc 4:79–94 (in Chinese, with English summary, Tables and Figures)

Chang CM (1969) Profile and clay mineral studies of brown forest soils in Taiwan. J Chin Agric Chem Soc 7:75–82 (in Chinese, with English summary, Tables and Figures)

Chang CM (1971) Clay mineralogy and related chemical properties of Podzolic soils in Taiwan. J Chin Agric Chem Soc 9:36–48 (in Chinese, with English summary, Tables and Figures)

Chang CM (1972) Mineralogy and related chemical properties of the soils developed from volcanic detritus and two pyroxene andesite in the Tatunshan region. Mem Coll Agric Natl Taiwan Univ 13(2):119–136 (in Chinese, with English summary, Tables and Figures)

Chang CM (1974) Mineralogical and related chemical properties of the soils developed from two pyroxene-hornblende andesite and olivine bearing hornblende-two pyroxene andesite in Tatunshan region. J Chin Agric Chem Soc 12:36–53 (in Chinese, with English summary, Tables and Figures)

Chang CM (1975) Clay mineralogy and related chemical properties of the soils developed from basalt and agglomerate in the Tatunshan region. J Chin Agric Chem Soc 13:46–57 (in Chinese, with English summary, Tables and Figures)

Chang JM, Chen ZS (1982) Morphology, physicochemical properties, genesis, and numerical classification of the seashore district soils of Taoyuan, Taiwan. J Agric Assoc China New Ser 119:1–24 (in Chinese, with English abstract, figures and tables)

Chang CM, Houng KH, Chen TT (1962) Profile and clay mineral studies of Taiwan soils. Bull Agric Chem Natl Taiwan Univ 11:22–43

Chang JM, Chen ZS, Chen CC, Lin CF (1983) Soil fertility characteristics of Taoyuan paddy soils and their significance in soil numerical classification: I. Alluvial soils in Kaohsiung-Pingtung region and latosols and lateritic alluvial soils in Taoyuan prefecture. J Agric Assoc China New Ser. 123:50–68 (in Chinese, with English abstract, figures and tables)

Chen ZS (1984) A model for the nowadays soil survey and classification of paddy soils of Taiwan: the study on the formation, genesis and classification of the seashore district paddy soils in Taoyuan, Taiwan, 357 pp. Ph.D. Dissertation, Graduate Institute of Agricultural Chemistry, National Taiwan University, Taipei, Taiwan (Advisor by Prof. JM Chang) (in Chinese, with English abstract, figures and tables)

Chen ZS (1998) Selecting the indicators to evaluate the soil quality of Taiwan soils and approaching the national level of sustainable soil management. In: FFTC/ASPAC (ed) Proceedings of international seminar on soil management for sustainable agriculture in the tropics, 14–19 Dec 1998, Taichung, Taiwan, ROC, pp 131–171 (Invited speaker)

Chen ZS, Chang JM (1984) Morphology, mineral composition and formation model of plinthite in the Taiwan soils. (I) Paddy soils of the seashore district of Taoyuan. Mem Coll Agric Natl Taiwan Univ 24:65–82 (in Chinese, with English abstract, figures and tables)

Chen ZS, Chang JM (1985) An investigation on the composition, distribution and their pedogenic characteristics of free oxides in well drainage paddy soils. Mem Coll Agric Natl Taiwan Univ 25:22–39 (in Chinese, with English abstract, figures and tables)

Chen ZS, Huang JH (1991) Characteristics and clay minerals of Andisols with melanic epipedon in Taiwan. J Chin Agric Chem Soc 29:415–426 (in Chinese, with English abstract, figures and tables)

Chen ZS, Tsai CC (2000) Morphological characteristics and classification of Podzolic soils in Taiwan. Soils Environ 3:49–62 (in Chinese, with English abstract, figures and tables)

Chen ZS, Lin KC, Chang JM (1989) Soil characteristics, pedogenesis, and classification of Bei-Cha-Tein-Shan Podzolic soils, Taiwan. J Chin Agric Chem Soc 27:145–155 (in Chinese, with English abstract, figures and tables)

Chen ZS, Liu JC, Chiang HC (1995) Soil properties, clay mineralogy, and genesis of some alpine forest soils in Ho-Huan Mountain area of Taiwan. J Chin Agric Chem Soc 33:1–17

Chen ZS, Asio VB, Yi DF (1999a) Characteristics and genesis of volcanic soils along a toposequence under a subtropical climate in Taiwan. Soil Sci 164:510–525

Chen ZS, Hseu ZY, Tsai CC (1999b) Total organic carbon pools in Taiwan rural soils and its application in sustainable soil management system. Soils Environ 1:295–306 (in Chinese, with English abstract, figures and tables)

Chen ZS, Tsou TC, Asio VB, Tsai CC (2001) Genesis of inceptisols on a volcanic landscape in Taiwan. Soil Sci 166:255–266

Chiang HC, Chen ZS, Hseu ZY (1997) Micromorphology of iron nodules in a montane Ultisols of central Taiwan. J Taiwan Sci 12:413–420 (in Chinese, with English abstract, figures and tables)

Delcaillau B, Deffontaines B, Floissac L, Angelier J, Deramond J, Souquet P, Chu HT, Lee JF (1998) Morphotectonic evidence from lateral propagation of an active frontal fold; Pakuashan anticline, foothills of Taiwan. Geomorphology 24:263–290

Guo HY, Liu TS, Chu CL, Chiang CF (1992) Introduction of soil information system in Taiwan. In: TARI Annual Reports 1992. Taichung, Taiwan, RROC (in Chinese)

Guo HY, Liu CH, Chu CL, Chiang CH, Yei MC (2000) The detailed grid soil survey of Taiwan rural soils. Taiwan Agricultural Research Institute, Council of Agriculture, Taichung (http://www.tari.gov.tw/index_in.htm)

Guo HY, Liu TS, Chu CL, Chiang CF, Yeh MJ (2002) Soil information of Taiwan and its development. In: Proceeding of Conference of soil information application, Taipei, Taiwan, pp 1–38 (in Chinese)

Guo HY, Liu CH, Chu CL, Chiang CH, Yei MC (2003) Report of Taiwan soil quality monitoring project (1998–2003). Taiwan Agricultural Research Institute, Council of Agriculture, Taichung

Guo HY, Liu CH, Chu CL, Chiang CH, Yei MC (2005) Taiwan soil information system. Taiwan Agricultural Research Institute, Council of Agriculture, Taichung (http://www.tari.gov.tw/index_in.htm)

Guo HY, Liu TS, Chu CL, Chiang CF, Römkens PF (2007) Prediction of heavy metal uptake by different rice species in paddy soils near contaminated sites of Taiwan. In: Proceeding of the 8th conference of East and Southeast Asian Federation of Soil Science Societies (ESAFS), Tsukuba, Japan

Harden JW (1982) A quantitative index of soil development from field descriptions, examples from a chronosequence in Central California. Geoderma 28:1–28

Ho CS (1988) An introduction to the geology of Taiwan explanatory text of the geologic map of Taiwan, 2nd edn. Central Geological Survey, Taipei, Taiwan, 164 pp

Hseu ZY, Chen ZS (1994) Micromorphology of rice-growing Alfisols with different wetness conditions in Taiwan. J Chin Agric Chem Soc 32:647–656

Hseu ZY, Chen ZS (1995) Redox processes of rice-growing Alfisols with different wet conditions. J Chin Agric Chem Soc 33:333–344

Hseu ZY, Chen ZS (1996) Saturation, reduction, and redox morphology of seasonally flooded Alfisols in Taiwan. Soil Sci Soc Am J 60:941–949

Hseu ZY, Chen ZS (1997) Soil hydrology and micromorphology of illuvial clay in an Ultisol hydrosequence. J Chin Agric Chem Soc 35:503–512

Hseu ZY, Chen ZS (1999) Micromorphology of redoximorphic features of subtropical anthraquic Ultisols. Food Sci Agric Chem 1:194–202

Hseu ZY, Chen ZS (2000) Monitoring changes of redox potential, pH, and electrical conductivity of the mangrove soils in northern Taiwan. Proc Natl Sci Counc ROC Part B Life Sci 24:143–150

Hseu ZY, Chen ZS (2001) Quantifying soil hydromorphology of a rice-growing Ultisol toposequence in Taiwan. Soil Sci Soc Am J 65:270–278

Hseu ZY, Chen ZS, Tsai CC (1999a) Selected indicators and conceptual framework for assessment methods of soil quality in arable soils of Taiwan. Soils Environ 2:77–88 (in Chinese, with English abstract, figures and tables)

Hseu ZY, Chen ZS, Wu ZD (1999b) Characterization of placic horizons in two subalpine Inceptisols. Soil Sci Soc Am J 63:941–947

Hseu ZY, Chen ZS, Leu IY (2000) Soil solution composition, water tables, and redox potentials of anthraquic Ultisols in a toposequence. Soil Sci 165:587–596

Hseu ZY, Tsai CC, Lin CW, Chen ZS (2004) Transitional soil characteristics between Ultisols and Spodosols in the Subalpine Forest of Taiwan. Soil Sci 169:457–467

Huang JH, Chen ZS (1990) Characteristics, genesis, and classification of two Andisols in Chihsingshan Mountain area, northern Taiwan. J Chin Agric Chem Soc 28:135–147 (in Chinese, with English abstract, figures and tables)

Huang JH, Chen ZS (1992) Soil properties and classification of Andisols in northeastern Chihsing mountain, Taiwan. J Chin Agric Chem Soc 30:216–228 (in Chinese, with English abstract, figures and tables)

Huang RH, Chang ZS, Chang JM (1988) Comparison of extraction for metals in Taiwan representative soils. J Chin Agric Chem Soc 26:476–485

Huang JH, Chen ZS, Wang MK (1993) Soil properties and clay mineralogy of Andisols in the northeastern Tatung mountain, Taiwan. J Chin Agric Chem Soc 31:325–339 (in Chinese, with English abstract, figures and tables)

Huang JH, Chen ZS, Wang MK (1994) The properties and classification of pedons derived from volcanic lava materials between Tatung mountain and Meinten mountain. J Chin Agric Chem Soc 32:294–308 (in Chinese, with English abstract, figures and tables)

Huang WS, Tsaia H, Tsai CC, Hseu ZY, Chen ZS (2010) Subtropical developmental soil sequence of Holocene marine terraces in eastern Taiwan. Soil Sci Soc Am J 74:1271–1283

Jien SH, Hseu ZY, Chen ZS (2004) Relations between morphological color index and soil wetness condition of anthraquic soils in Taiwan. Soil Sci 169:871–882

Lee JC, Lu CY, Chu HT, Delcaillau B, Angelier J, Deffontaines B (1996) Active deformation and paleostress analysis in the Pakua anticline area, western Taiwan. TAO 7:431–446

Lee CH, Wu MY, Asio VB, Chen ZS (2006) Using soil quality index to assess the effects of applying manure compost on the soil quality under a crop rotation system in Taiwan. Soil Sci (in press) (SCI)

Li SY, Chen ZS, Liu JC (1998a) Subalpine loamy Spodosols in central Taiwan: Characteristics, micromorphology, and genesis. Soil Sci Soc Am J 62:710–716

Li SY, Chen ZS, Liu JC (1998b) Weathering sequence of clay minerals in loamy Spodosols of Central Taiwan. J Agric Assoc China New Ser 183:51–68

Lin KC, Chen ZS, Chang JM (1988) Morphology, physicochemical properties, and classification of Lalashan Podzolic soils of northern Taiwan. J Chin Agric Chem Soc 26:503–514 (in Chinese, with English abstract, figures and tables)

Lin CW, Hseu ZY, Chen ZS (2002) Clay Mineralogy of Spodosols with high clay contents in the Subalpine forests of Taiwan. Clays Clay Miner 50(6):726–735

Liu JT, Chen ZS (1990) Characteristics, genesis, and classification of podzolic soils in Tamanshan Mountain, northern Taiwan. J Chin Agric Chem Soc 28:148–159 (in Chinese, with English abstract, figures and tables)

Liu JT, Chen ZS (1991) Lepidocrocite and goethite in the forest soils of the Taman mountains, Taiwan. J Chin Agric Chem Soc 29:484–491 (in Chinese, with English abstract, figures and tables)

Liu JC, Chen ZS (1998) Soil characteristics and pedogenesis of loamy Spodosols near Wanghsiang mountain in central Taiwan. J Chin Agric Chem Soc 36:151–162

Liu JC, Chen ZS (2004) Soil characteristics and clay mineralogy of two Spodosols in central Taiwan. Soil Sci 169:66–80

Liu JC, Chen ZS, White GN, Dixon JB (1998) Goethite in an Alfisol and an Ultisol of southern Taiwan. J Agric Assoc China New Ser 182:83–98

Saigusa M, Matsuyama N, Honna T, Abe T (1991) Chemistry and fertility of acid Andisols with special reference to subsoil acidity. In: Wright RJ et al (eds) Plant-soil interactions at low pH. Kluwer Academic Publishers, Netherlands, pp 73–80

Shoji S, Fujiwara YS, Yamada I, Saigusa M (1982) Chemistry and clay mineralogy of ando soils, brown forest soils, and podzolic soils formed from recent Towada ashes, northeastern Japan. Soil Sci 133:69–85 (SCI)

Shoji S, Nanzyo M, Dahlgren R (1993) Volcanic ash soils: Genesis, properties and utilization. Development in Soil Science #21. Elsevier Publisher, Amsterdam, London, New York, and Tokyo. ISBN: 0-444-89799-2

Si LC, Chang SC (1951) Report of soil survey and soil fertility of Taiwan. Book series No.5, Taiwan Fertilizer Company, Taipei, Taiwan (in Chinese)

Soil Survey Staff (1998) Keys to soil taxonomy, 8th edn. U.S. Department of Agriculture, Soil Conservation Service, Washington, DC

Soil Survey Staff (1999) Soil Taxonomy: a basic system of soil classification for making and interpreting soil surveys. In: USDA, Natural resources conservation service, agriculture. Handbook No. 436. 2nd edn. Washington, DC

Soil Survey Staff (2003) Keys to soil taxonomy, 9th edn. U.S. Department of Agriculture, Soil Conservation Service, Washington, DC

TFRI (Taiwan Forest Research Institute, Council of Agriculture) (2005) Soil survey report, soil maps, and soil database of Taiwan forest soils (http://ims.tfri.gov.tw/tfri1/main.aspx/)

Tsai CC, Chen ZS (2006) Soil mapping by terrain analysis and geographic information systems: a case study of Taiwan volcanic soils (unpublished)

Tsai H, Sung QC (2003) Geomorphic evidence for an active pop-up zone associated with the Chelungpu fault in central Taiwan. Geomorphology 56:31–47

Tsai H, Huang WS, Hseu ZY, Chen ZS (2006) A river terrace soil chronosequence of the Pakua tableland in central Taiwan. Soil Sci 171:167–179

Tsai CC, Tsai H, Hseu ZY, Chen ZS (2007a) Soil pedogenesis along a chronosequence on marine terraces in eastern Taiwan. Catena 71:394–405

Tsai H, Hseu ZY, Huang WS, Chen ZS (2007b) Pedogenic approach to the geomorphic study on river terraces of the Pakua tableland in Taiwan. Geomorphology 83:14–28

Tsai H, Huang WS, Hseu ZY (2007c) Pedogenic correlation of the lateritic river terraces in central Taiwan. Geomorphology 88:201–213

Tsai H, Maejima Y, Hseu ZY (2008) Meteoric [10]Be dating of highly weathered soils from fluvial terraces in Taiwan. Quatern Int 188:185–196

Tsai H, Hseu ZY, Huang WS, Huang ST, Chen ZS (2010) Pedogenic properties of surface deposit as evidences on landform formation of the Tadu tableland in central Taiwan. Geomorphology 114:590–600

Wu XP, Chen ZS (2005) Soil characteristics and genesis of Inceptisols with placic horizon in subalpine forest soils of Taiwan. Geoderma 125:331–341

3.1 General Information

Alfisols are characterized by an argillic horizon in which silicate clay has accumulated by illuviation due to the large amount of annual rainfall in Taiwan. Clear clay coatings are present on the walls of pores or on the pedsurfaces in such a subsurface diagnostic horizon. However, this clay accumulated horizon is moderately leached, and thus its cation exchange capacity is more than 35 % base saturated and corresponds to neutral to weak alkaline in pH regimes.

In general, Alfisols are locally found in aged alluvial plains with sandstone and shale sediments in southwestern Taiwan. Additionally, a few Alfisols are inherited from the moderate weathering of mafic or ultramafic rocks, such as basalt in Penghu Island, uplifted coral reef in southern tips of Taiwan, and serpentine in Eastern Taiwan. However, the Alfisols are well- and excessive-drained except for some of them in the aged alluvial plains for paddy production.

Three groups in Soil Taxonomy are currently found in the Alfisols of Taiwan. They are Aqualfs, Udalfs, and Ustalfs. The distribution pattern of these soil types closely follows human activity and climate. The Aqualfs are considered to be transformed from Ultisols by human activities. Certain Ultisols have been long-term fertilized for paddy production to increase the amounts of base cations (e.g., Ca, K) adsorbed on the surfaces of soil particles, so that the increase of base saturation results in the soils to become Alfisols (Hseu and Chen 1996). Taiwan is a humid and tropical/subtropical country, but the unique difference between the Udalfs and Ustalfs is soil moisture regime depending on local strength of seasonal monsoon from China. In Chiayi and Tainan, southwestern Taiwan, high evaporation rates associated with relatively low rainfall produced by the strong northeast monsoon in winter, and thus the Ustalfs were formed. The other Alfisols are generally classified as Udalfs, which are more moisturized by humid climate and forest vegetation in comparison with the Ustalfs. The Udalfs are dominant great groups of Alfisols in Taiwan.

3.2 Profile Characteristics

3.2.1 Serpentine-Derived Soil

Serpentine is a product of hydrothermally metamorphic ultramafic rocks such as peridotites, dunites, and pyroxenites (Fig. 3.1). Although serpentines and soils cover only a small area of the Earth's terrestrial surface, they are abundant in ophiolite belts and have been typically found within regions of the Circum-Pacific margin and Mediterranean Sea (Oze et al. 2004a, b). Exposed serpentines on the surface are locally found along the eastern sides of the Central Range and the western sides of the Coastal Range, Eastern Taiwan. Serpentines alteration as well as time affected the pedogenesis. However, Alfisols are one of the serpentine-derived soils, which are in the moderate stage of soil development.

In morphology, the Alfisols from serpentines are deep. A typical Alfisol (TA pedon) from serpentines is illustrated in this chapter, which is located in the Tong-An Shan area of Coastal Range in Eastern Taiwan. The soil moisture regime is udic, and the soil temperature regime is hyperthermic in the surrounding area, in which subtropical broad-leaved evergreen forests are the dominant vegetation. The TA pedon is located in the suture zone (Longitudinal Valley in Eastern Taiwan) of the arc-continent collision (Ho 1988). The parent material of Pedon is the residuum of exotic blocks of serpentinitic rocks underlain by the mudstone matrix of the Lichi Formation during the Pliocene and Pleistocene. This pedon is on the flat landform (Fig. 3.2). As expected in TA pedon with flat landscape, the soil has generally undergone intensive weathering, extensive leaching, causing metal, and clay translocation.

Field morphology of the TA pedon is described as follows (Fig. 3.3). If not otherwise indicated, soil color is given for moist soil.

Z.-S. Chen et al., *The Soils of Taiwan*, World Soils Book Series,
DOI 10.1007/978-94-017-9726-9_3, © Springer Science+Business Media Dordrecht 2015

Ae	0–17 cm	Very dark gray (7.5YR 3/1) silty clay loam; moderate medium and coarse subangular blocky structures; firm, slight sticky and slight plastic; many fine and medium roots; clear wavy boundary
Bt1	17–48 cm	Dark brown (7.5YR 3/2) clay; strong medium and coarse subangular blocky structures; firm, slight sticky and plastic; discontinuous few cutans on the pedsurface; many very fine and finer roots; gradual wavy boundary
Bt2	48–72 cm	Dark reddish brown (5YR 3/3) clay; strong medium and coarse subangular blocky structures; firm, sticky and plastic; continuous some cutans on the pedsurface; many very fine and fine roots; gradual wavy boundary
Bt3	72–101 cm	Dark reddish brown (5YR 3/3) clay; strong medium and coarse subangular blocky structures; firm, sticky and plastic; continuous some cutans on the pedsurface; many very fine and fine roots; diffuse wavy boundary
Bt4	101–140 cm	Brown (7.5YR 4/3) clay; moderate medium and coarse subangular blocky structures; firm, sticky and plastic; continuous some cutans on the pedsurface; few very fine and fine roots; diffuse wavy boundary
Bt5	140–170 cm	Brown (7.5YR 4/3) clay; moderate medium and coarse subangular blocky structures; firm, sticky and plastic; continuous some cutans on the pedsurface; few very fine roots; diffuse wavy boundary
Bt6	170–200 cm	Yellowish brown (10YR 5/4) clay loam; moderate medium and coarse subangular blocky structures; firm, sticky and plastic; discontinuous few cutans on the pedsurface; diffuse wavy boundary
BC1	200–240 cm	Light yellowish brown (10YR 6/4) loam; moderate medium subangular blocky structures; firm, sticky and plastic; diffuse wavy boundary
BC2	240–280 cm	Light yellowish brown (10YR 6/4) loam; moderate medium subangular blocky structures; firm, sticky and plastic; diffuse smooth boundary
C	>280 cm	Olive brown (2.5Y 4/4) loam; weak fine and medium subangular blocky structures; friable, slightly sticky and plastic

Fig. 3.1 Serpentine outcrop in a field of Eastern Taiwan. Copyright by Prof. Hseu

Fig. 3.2 Alfisol landscape on serpentines in Eastern Taiwan. Copyright by Prof. Hseu

In micromorphology, striated clay coatings on the pedsurface are very clear with strong orientation and highly birefringence in the illuvial horizons of TA pedon (Fig. 3.4). Additionally, Fe and Mn concentrated materials as black or dark red pellets in the soil matrix of the illuvial horizons were elucidated the metal migration through the profile.

Clear illuviation is identified in TA pedon based on clay coatings in the field and by differences in particle size analysis data. The C horizon has a loamy texture with >20 % clay, and all the B horizons are clayey textured (Table 3.1). The concentrations of DCB-extractable Fe (Fe_d) are generally higher in the Bt horizon of the profile. The accumulation of clay and Fe_d in the subsurface horizon resulted mainly from the strong leaching potential owing to extremely high rainfall. The pH values of all horizons were below 7.0, and generally increased with soil depth in the pedon. This pedon derived from ultramafic rocks are characterized by relatively high cation exchange capacity, ranging from approximately 34–52.2 cmol/kg in all horizons.

The exchangeable Ca/Mg ratios were below 1.0, with the only exceptions being the surface horizons. Alexander et al. (1989) and Lee et al. (2004) indicated that highly developed serpentinitic soils displayed higher Ca/Mg ratios because Mg was leached out of the soils and plant detritus containing Ca was added to the soils. Therefore, the relatively highly developed soils of TA pendon exhibited relatively low

organic horizon and decrease with soil depth. Moreover, the OC content was always below 3.0 %.

Soils derived from serpentinites may display strong chemical fertility limitations owing both to low Ca/Mg ratio and to non-anthropogenic Cr and Ni enrichments (Bonifacio et al. 1997; Burt et al. 2001; Hseu 2006; Cheng et al. 2009, 2011). Chromium content in non-serpentine soils and rocks generally ranges up to 200 mg/kg (Alloway 1991). However, Cr contents of up to 10,000 mg/kg have been reported in soils developed on ultramafic rocks (Alexander et al. 2007). The forms in which Cr and Ni exist in serpentinitic soils are fundamental for predicting their bioavailability and mobility in soil environments. The BCR sequential extraction was valid to explore Cr and Ni fractions in the TA pedon by Hseu (2006), which divided Cr and Ni into the following fractions: (1) acid extractable (F1), (2) reducible (F2), (3) oxidizable (F3), and (F4) residual (namely silicate mineral-bound, F4) (Quevauviller et al. 1994). In Table 3.2, the most mobile phase in the soils, F1 fraction, was significantly lower than the other fractions in the soils. The trends of F2 and F3 resemble that of F1. In the pedon, levels of four fractions always displayed the following order: F4 > F2 > F3 > F1.

Nickel was primarily recovered in the F4 compartments because large parts of Ni were primarily fixed in silicate lattices such as serpentine (Table 3.3). Relatively low concentrations of Ni existed in the acid extractable and oxidizable forms. Nevertheless, the F1 and F2 fractions of Ni generally exceed those of Cr. In comparison with the proportions of F1 and F2 in total content of metals between Cr and Ni, Ni was more mobile than Cr in the serpentine soils. These proportions of Ni increased with decreasing soil depth, indicating that Ni was gradually transformed into easily mobile phases during the weathering of serpentinitic rocks. This result agrees with the study of Gouch et al. (1989).

Fig. 3.3 Field morphological characteristics of TA pedon. Copyright by Prof. Hseu

Ca/Mg values. The base saturation (BS) of the diagnostic horizon in TA pedon generally exceeds 35 %. Organic C (OC) contents are generally the greatest in the surface

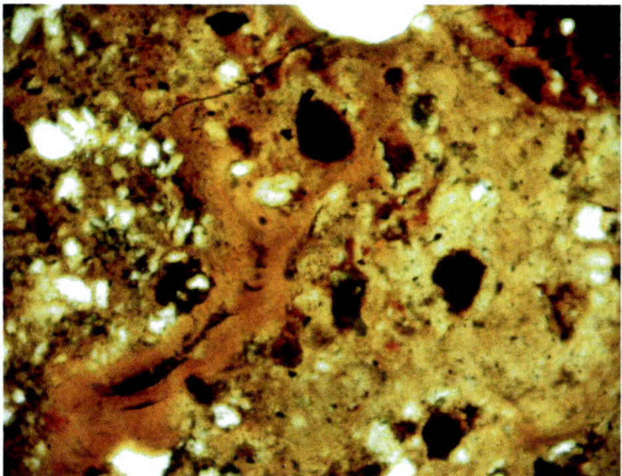
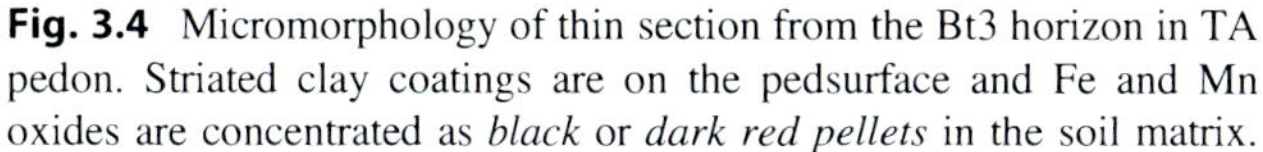
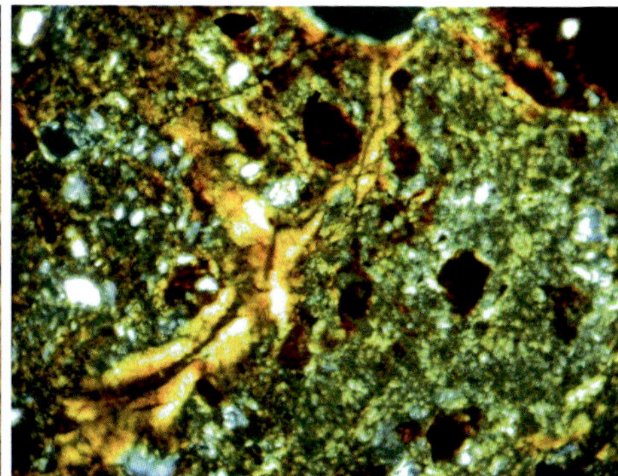

Fig. 3.4 Micromorphology of thin section from the Bt3 horizon in TA pedon. Striated clay coatings are on the pedsurface and Fe and Mn oxides are concentrated as *black* or *dark red pellets* in the soil matrix.

Left photo is under plane polarized light (PPL) and *right photo* is under cross polarized light (XPL). The width of each frame is 0.625 mm. Copyright by Prof. Hseu

Table 3.1 Selected properties of the TA pedon on serpentines from Eastern Taiwan

Horizon	Depth (cm)	Sand (%)	Silt (%)	Clay (%)	OC^a (%)	pH	BS^b (%)	CEC^c (cmol/kg)	Fe_d^d (g/kg)	Ca/Mg^e
Ae	0–19	16	49	35	2.4	5.2	37	34.7	32.1	1.1
Bt1	19–48	8	40	53	1.2	5.1	44	41.9	34.5	1.1
Bt2	48–72	13	17	70	1.5	5.1	46	41.5	43.8	0.9
Bt3	72–101	12	23	65	0.8	5.3	45	50.9	45.5	0.9
Bt4	101–140	19	31	50	0.4	5.9	56	50.5	33.8	0.5
Bt5	140–170	22	32	47	0.6	6.1	57	49.7	22.7	0.8
Bt6	170–200	37	31	32	0.3	6.2	55	53.8	16.3	0.8
BC1	200–240	47	28	25	0.7	6.6	52	42.7	6.73	0.8
BC2	240–280	41	33	26	0.6	6.7	53	52.5	6.22	0.5
C	<280	48	31	21	0.6	6.7	50	42.9	5.90	0.6

[a] Organic carbon
[b] Percentage of base saturation
[c] Cation exchange capacity
[d] DCB-extractable Fe
[e] Ca and Mg are exchangeable form
Modified from Hseu (2006)

Table 3.2 Different fractions of Cr in the Alfisol pedon on serpentinites from Eastern Taiwan

Horizon	Depth (cm)	Sequential extraction				Σ^a	Total content (g/kg)	Recovery (%)
		F1 (mg/kg)	F2 (mg/kg)	F3 (mg/kg)	F4 (mg/kg)			
Ae	0–19	0.57	46.3	2.56	228	277	0.36	77.7
Bt1	19–48	0.81	62.1	2.06	245	310	0.36	87.0
Bt2	48–72	0.38	97.2	2.15	277	377	0.36	105
Bt3	72–101	0.28	108	3.35	324	435	0.42	103
Bt4	101–140	ND^b	85.4	4.55	273	363	0.34	108
Bt5	140–170	ND	75.6	4.95	209	289	0.27	108
Bt6	170–200	ND	49.6	3.50	166	219	0.21	107
BC1	200–240	1.86	14.8	2.58	228	247	0.20	123
BC2	240–280	0.81	13.8	2.93	150	168	0.24	71.2
C	>280	1.87	25.1	2.66	170	199	0.27	74.3

[a] Sum of heavy metals content in each fraction
[b] Not detectable
Modified from Hseu (2006)

Table 3.3 Different fractions of Ni in the Alfisol pedon on serpentinites from Eastern Taiwan

Horizon	Depth (cm)	Sequential extraction				Total content (g/kg)	$Recovery^a$ (%)
		F1 (mg/kg)	F2 (mg/kg)	F3 (mg/kg)	F4 (mg/kg)		
Ae	0–19	5.83	34.7	17.3	208	0.34	77.9
Bt1	19–48	5.08	34.0	5.52	228	0.29	92.7
Bt2	48–72	0.62	29.8	7.23	290	0.38	85.3
Bt3	72–101	2.32	30.1	6.04	288	0.36	90.1
Bt4	101–140	1.57	58.9	7.19	192	0.27	96.7
Bt5	140–170	1.05	41.3	2.33	178	0.23	96.7
Bt6	170–200	1.71	51.2	0.60	140	0.19	101
BC1	200–240	ND^b	2.01	0.99	161	0.16	104
BC2	240–280	ND	1.98	1.37	172	0.17	102
C	>280	ND	13.2	2.08	161	0.16	113

[a] Sum of heavy metals content in each fraction/total content
[b] Not detectable
Modified from Hseu (2006)

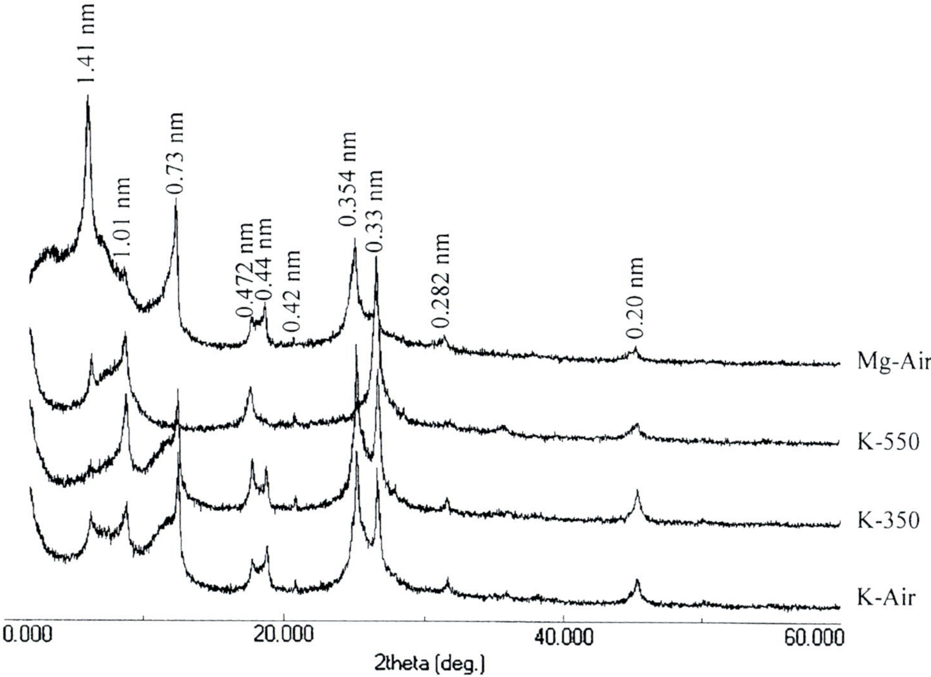

Fig. 3.5 XRD of clay fraction in the Bt5 horizon of TA pedon. Copyright by Prof. Hseu

Hseu et al. (2007) selected four pedons along a toposequence from the summit (Entisol), shoulder (Vertisol), backslope (Alfisol, the TA pedon) to footslope (Ultisol) positions to explore the contributions and the significance of landscape and weathering status of serpentinitic rock with regard to clay mineral transformations in Eastern Taiwan.

The XRD patterns of clay fractions in the Bt5 horizon of the TA pedon are illustrated to explain the clay mineralogy (Fig. 3.5). The reflection at 25 °C in the K-saturated clays was present at 0.73 nm, and it could contain chlorite attributing that the peak was reduced when heated to 550 °C. On the other hand, chlorite was clearly characterized by the peaks at 1.41, 0.73, 0.472, and 0.354-nm. Chlorite was present in all horizons of the pedon, however, it increased with soil depth. Vermiculite was characterized both by the peak at 1.0-nm when the K-saturated clays were heated at 110 °C and by the peak at 1.41-nm when the Mg-saturated clays were solvated with glycerol. The basal XRD peaks at 1.41-nm with the Mg-saturated clays and intermediated between 1.2-nm and 1.3-nm with the K-saturated clays heated at 550 °C are the characteristics of randomly interstratified chlorite-vermiculite.

The clay mineralogy reflects the clear weathering trend of the soils along the toposequence: (1) The soils on the summit and shoulder contained smectite and serpentine, which are predominant in the young soils derived from serpentinitic rocks, and (2) vermiculite gradually increased in the relatively old soils on backslope and footslope. The mineralogical transformations observed along the toposequence indicated that chlorite and serpentine, initially present in the Entisol on the summit, weather into smectite and interstratified chlorite-vermiculite in the intermediate soil on the shoulder under strong leaching and oxidizing conditions. Furthermore, vermiculite formed as the major weathering product of chlorite and smectite in the soil developed on the backslope.

3.2.2 Taiwan Clay

"Taiwan clay" is the popular name given by farmers, which is locally found on the aged alluvial plain of sand and shale in southwestern Taiwan. In Soil Taxonomy, Taiwan clay is generally classified as Alfisols and characterized with (1) high soil depth (>150 cm), (2) well or moderately well drained, (3) the color is commonly gray or light olive in the surface soil with many yellowish brown to yellowish red mottles; the color is gray, dark gray, brownish gray, or olive in the subsurface soil, (4) fine textured dominated by silty clay loam or silty clay, (5) the coarse column or angular blocky soil structures coated by gray clay, (6) the soil consistence is very firm, (7) the soil reaction is weak or moderate alkaline, and (8) there is no carbonate nodule through soil profile with lime reaction.

Field morphology of the Shiu-yuo pedon (Typic Haplustalf) is described as follows (Fig. 3.6). If not otherwise indicated, soil color is given for moist soil.

Fig. 3.6 Taiwan Clay pedon in Tainan county, Southern Taiwan. Copyright by Prof. Chen

Ap1	0–15 cm	Gray (5Y 6/1) with reddish yellow (7.5YR 6/8) mottle silty clay; strong medium subangular blocky structures; firm, sticky and plastic; few fine and medium roots; clear smooth boundary
Ap2	15–38 cm	Dark gray (5Y 4/1) silty clay; strong coarse subangular blocky structures; very firm, sticky and plastic; few fine roots; clear wavy boundary
Bt1	38–76 cm	Dark gray (5Y 4/1) silty clay; strong coarse column structures; very firm, very sticky and plastic; few clear discontinuous clay coatings on the pedsurfaces; few very fine roots; diffuse smooth boundary
Bt2	>76 cm	Light olive brown (2.5Y 5/4) silty clay loam; moderate medium subangular blocky structures; very firm, very sticky and plastic; some clear continuous clay coatings on the pedsurfaces; few very fine roots; diffuse smooth boundary

Except for the surface layer, clay content the Taiwan clay is high, particularly in the subsurface soils with clay content over 30 % (Table 3.4). The climate of the Taiwan clay area is characterized by high air temperature and relatively low rainfall, compared to other pedoclimatic areas of Taiwan. Therefore, the soil pH of the Taiwan clay is high and correspond to high base saturation. Additionally, free oxides of Fe, Al, and Mn are clearly accumulated in the B horizons.

3.2.3 Coral Reef-Derived Soil

Taiwan is suited in the northeast trending collision zone of the Eurasian plate and Philippine Sea plate. The active tectonics result in uplifted coral reefs, such as the Kengting Uplifted Coral Reef Nature Reserve (KUCRNR) in Southern Taiwan (Ho 1988). This ecosystem, a tropical lowland evergreen broad-leaved rain forest, is unique because of uplifted coral reef landscape and vegetation.

The KUCRNR, with an area of approximately 138 ha, is located on Hengchun Peninsula, at the southern tip of Taiwan Island. This nature reserve is the only fully conserved forest underlain primarily by Miocene rocks and capped with uplifted coral reef and recent sediments in Taiwan. The underlying formation is Hengchun Limestone, which is an organic reef comprising the remains of various organisms, including corals, foraminifers, mollusks, and calcareous algae. These organisms are more abundant in the lower part of the limestone unit than in the higher part. When the limestone is well developed, it is gray to creamy white, and either massive and compact or porous. The Hengchun Limestone is middle to late Pleistocene. Consequently, the coral reef and recent sediments were uplifted in Hengchun Peninsula to form the present landscape. The elevation ranges from 200 to 300 m above sea level and slopes downward toward the south. In terms of altitude, the exposed coral reef is generally 10–20 m higher than the sediment basin or valley (Fig. 3.7).

Ebenaceae and *Euphorbiaceae* are the main families in the study area. Coast persimmon (*Diospyros maritime*) was the dominant species, and had relatively high density and dominance of 50.3 and 23.2 %, respectively. However, the common species are white bark fig (*Ficus benjamina*) and swamp gelonium (*Melanolepis multiglandulosa*), in addition to coast persimmon.

In the sediment basin between exposed coral reefs, Alfisols are commonly found in the reserve. Field morphology of the CR3 pedon is described as follows (Fig. 3.8). If not otherwise indicated, soil color is given for moist soil.

The surface soil textures of the four habitats are clayey textured with >40 % clay. The soil reactions were weakly acidic, and thus the $CaCO_3$ content ranged in 3.7–4.4 % and had high exchangeable Ca (>7.0 cmol/kg). The abundant Ca in the soils was primarily released from the weathering processes of coral reef (Table 3.5).

Horizon	Depth	Description
Ae	0–7 cm	Dark brown (7.5YR 3/2) sandy clay loam; moderate very fine and fine granular structures; friable, slight sticky and slight plastic; many fine and medium roots; gradual wavy boundary
BA	7–18 cm	Dark brown (7.5YR 3/3) clay loam; moderate fine and medium subangular blocky structures; friable, slight sticky and slight plastic; many fine roots; diffuse wavy boundary
Bt1	18–40 cm	Strong brown (7.5YR 4/6) clay loam; strong medium and coarse angular blocky structures; faint discontinuous clay coatings on the pedsurfaces; firm, sticky and plastic; some fine roots; diffuse smooth boundary
Bt2	40–70 cm	Brown (7.5YR 4/4) clay; strong medium and coarse angular blocky structures; clear continuous clay coatings on the pedsurfaces (Fig. 3.9); firm, sticky and plastic; some fine roots; diffuse smooth boundary
Bt3	70–100 cm	Brown (7.5YR 5/4) clay; strong medium and coarse angular blocky structures; clear continuous clay coatings on the pedsurfaces and root channels; firm, sticky and plastic; some fine roots; diffuse smooth boundary
Bt4	100–130 cm	Strong brown (7.5YR 5/6) clay; strong medium and coarse angular blocky structures; clear continuous clay coatings on the pedsurfaces and root channels; very firm, sticky and plastic; few fine roots; diffuse smooth boundary
Bt4	130–150 cm	Strong brown (7.5YR 5/6) clay; strong medium and coarse angular blocky structures; clear continuous clay coatings on the pedsurfaces; very firm, sticky and plastic; few fine roots; diffuse smooth boundary
Bt5	150–200 cm	Strong brown (7.5YR 5/6) clay; strong medium and coarse angular blocky structures; clear continuous clay coatings on the pedsurfaces; very firm, sticky and plastic; few very fine roots; diffuse smooth boundary

Annual litterfall in the study habitats ranged from 6.98 to 9.13 Mg/ha/year, and thus was in the higher order of the range for humid tropical forests (Liao et al. 2006). Litterfall was generally higher in habitats I and IV than in habitats II and III. In each habitat, litterfall displayed seasonal peaks through the year in June–August owing to frequent typhoons and heavy rainfall and again in December due to winter monsoon. The mass loss rates of leaf litter did not differ significantly between habitats, but was considerably higher than those in ecosystems of temperate regions or in non-coral reef tropical ecosystems. The fastest rate of decay of carbon in leaf litter occurred in habitat IV, yet the leaves of the dominant species, pisonia tree, are thin and lack prominent skeletal tissues. In all habitats, the mass loss percentages of Ca, Mg, K, and Na in leaf litter significantly exceeded those of C, N, and P, particularly at the end of the decomposition period, and thus, based on the extent of release, mass and nutrient mobility was ranked as follows: Ca > Mg = N > Na > K > P > mass > C. The quantities of nutrients in leaf litter returned to the soils were as follows: C > Ca > N > K > Mg > Na > P in the study area, and the sequence differed from the quantity sequence of these nutrients in the soils.

3.3 Formation and Genesis

In thickness, the Alfisols are generally deep (>150 cm). Clear illuviation is clearly identified in the Alfisols of Taiwan by field morphology, micromorphology, and particle size distribution of profile. Clay coating was generally formed in their argillic horizons as clear orientation under polarized microscopy. The high base saturation is resulted from parent materials in serpentine and coral reef areas and from high evaporation in the Taiwan clay area. Although eluvial/illuvial processes of clay and pedogenic Fe oxides are evident in the Alfisols, the amount of base cation cations in their subsurface horizons is still high particularly in serpentine-derived soils with more exchangeable Mg than Ca. However, the coral-reef soils and Taiwan clays have higher

Table 3.4 Selected properties of a Taiwan clay Alfisol

Horizon	Depth (cm)	Sand (%)	Silt (%)	Clay (%)	pH (%)	OC[a] (g/kg)	CEC[b] (cmol/kg)	BS[c] (%)	Free oxides		
									Fe (g/kg)	Al (g/kg)	Mn (g/kg)
Ap1	0–15	27	64	19	6.7	14.4	10.8	56	14.9	1.18	0.09
Ap2	15–38	12	53	35	7.7	4.05	12.8	67	20.2	2.06	0.32
Bt1	38–76	17	45	38	7.7	3.37	13.9	68	22.2	2.33	0.29
Bt2	>76	14	56	30	7.7	2.87	10.8	70	17.5	2.50	0.26

[a] Organic carbon
[b] Cation exchange capacity
[c] Base saturation
Copyright by Prof. Hseu

Fig. 3.7 Outcrop of coral reef in Hengchun Peninsula of southern tip of Taiwan. Copyright by Prof. Hseu

exchangeable Ca than Mg. The average OC content is not high in the Alfisols, in spite of the fact that the soils develop in tropical forest, such as the coral-reef soils, where runoff is observable in humid climate.

3.4 Land Use and Management

Serpentine soils often pose ecological or environmental risk due to high levels of potentially toxic metals such as Co, Cr, Mn, and Ni (Oze et al. 2004a, b; Hseu 2006), low concentrations of P and K (Becquer et al. 2003), low extractable Ca/Mg ratios (Alexander et al. 1989; Bonifacio et al. 1997; Lee et al. 2001), unique flora, and unique physical properties (Oze et al. 2008). Most serpentine soils in Taiwan are located in forests which are generally sparse on the landscape. Due to the limited fertility, it is rare in this area for crop production in Alfisols from serpentines where only paddy rice and corn are grown in limited places in Eastern Taiwan.

Fig. 3.8 Field morphology of Pedon CR3 at the Kengting Uplifted Coral Reef Nature Reserve (KUCRNR) in Hengchun Peninsula of Southern Taiwan. Copyright by Prof. Hseu

The major constraint in crop production is limited water availability for Taiwan Clay Alfisols, despite the pH regime and base cation concentrations being suitable for agricultural production. After the large-scale irrigation system ended in the 1930s, the alluvial soils on Chia-nan Plan became the most productive for agriculture in Taiwan including the Alfisols, particularly for staple crops like rice and corn. Regarding paddy rice on the Taiwan Clay Alfisols, the annual harvest generally exceeds 10 ton/ha.

The coral-reef soils are generally located in the ecologically preserved areas announced by the government, so that natural vegetation covers these soils and no clear human activities occur on the soils.

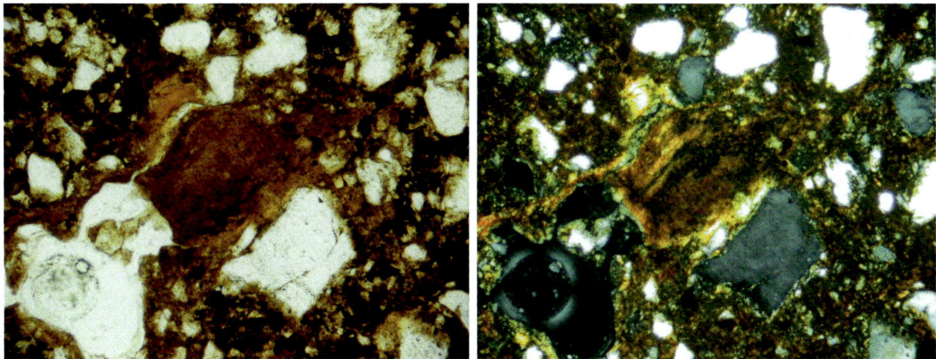

Fig. 3.9 Micromorphology of Bt2 horizon in Pedon CR3. Illuvial clays accumulated as dark brown infillings are on the pedsurface and in the chambers under PPL (*left*) which show strong birefringence and under XPL. The width of each frame is 0.625 mm. Copyright by Prof. Hseu

Table 3.5 Selected properties of the CR3 pedon on the KUCRNR in Hengchun Peninsula of Southern Taiwan

Horizon	Depth (cm)	Sand	Silt	Clay	Bd[a] (Mg/m^3)	pH	OC[b] (g/kg)	CEC[c] (cmol/kg)	BS[d] (%)	Carbonate (%)
Ae	0–7	58	14	28	1.2	6.2	1.66	17.5	100	1.12
BA	7–18	38	27	35	1.2	5.9	1.53	11.1	100	0.97
Bt1	18–40	44	22	54	1.2	5.6	0.71	9.86	100	0.58
Bt2	40–70	32	18	60	1.4	5.6	0.39	10.9	100	0.45
Bt3	70–100	24	16	60	1.6	5.7	0.35	12.7	100	0.48
Bt4	100–130	20	19	61	1.5	5.7	0.39	13.3	100	0.55
Bt5	130–150	17	10	63	1.5	5.8	0.39	14.8	87.1	0.46
Bt6	150–200	13	11	64	1.5	5.7	0.2	13.4	100	0.26

[a] Bulk density
[b] Organic carbon
[c] Cation exchange capacity
[d] Base saturation percentage
Copyright by Prof. Hseu

References

Alexander EB, Adamson C, Graham RC, Zinke PJ (1989) Soils and conifer forest productivity on serpentinized peridotite of the trinity ophiolite, California. Soil Sci 148:412–423

Alexander EB, Coleman RG, Keeler-Wolf T, Harrison S (2007) Serpentine geoecology of western North America. Oxford University Press, New York (USA), 512 pp

Alloway BJ (1991) Heavy metals in soils. Blackie and Son Ltd., London, 339 pp

Becquer T, Quantin C, Sicot M, Boudot JP (2003) Chromium availability in ultramafic soils from New Caledonia. Sci Total Environ 301:251–261

Bonifacio E, Zanini E, Boero V, Franchini-Angela M (1997) Pedogenesis in a soil catena on serpentinite in north-western Italy. Geoderma 75:33–51

Burt R, Fillmore M, Wilson MA, Gross ER, Langridge RW, Lammers DA (2001) Soil properties of selected pedons on ultramafic rocks in Klamath Mountains, Oregon. Commun Soil Sci Plant Anal 32:2145–2175

Cheng CH, Jien SH, Tsai H, Chang YH, Chen YC, Hseu ZY (2009) Geochemical element differentiation in serpentine soils from the ophiolite complexes, Eastern Taiwan. Soil Sci 174:283–291

Cheng CH, Jien SH, Iizuka Y, Tsai H, Chang YS, Hseu ZY (2011) Pedogenic chromium and nickel partitioning in serpentine soils along a toposequence. Soil Sci Soc Am J 75:659–668

Gouch LP, Meadows GR, Jackson LL, Dudka S (1989) Biogeochemistry of a highly serpentinized, chromite rich ultramafic area, Tehama County, California. USGS Bulletin No. 1901, US Dept. of the Interior, Washington, DC

Ho CS (1988) An introduction to the geology of Taiwan: explanatory text of the geologic map of Taiwan, 2nd edn. Central Geological Survey, Taiwan, 192 pp

Hseu ZY (2006) Concentration and distribution of chromium and nickel fractions along a serpentinitic toposequence. Soil Sci 171:341–353

Hseu ZY, Chen ZS (1996) Saturation, reduction, and redox morphology of seasonally flooded Alfisols in Taiwan. Soil Sci Soc Am J 60: 941–949

Hseu ZY, Tsai H, Hsi HC, Chen YC (2007) Weathering sequences of clay minerals in soils along a serpentinitic toposequence. Clays Clay Miner 55:389–401

Lee BD, Graham RC, Laurent TE, Amrhein C, Creasy RM (2001) Spatial distribution of soil chemical conditions in a serpentinitic wetland and surrounding landscape. Soil Sci Soc Am J 65:1183–1196

Lee BD, Graham RC, Laurent TE, Amrhein C (2004) Pedogenesis in a wetland meadow and surrounding serpentinic landslide terrain, northern California, USA. Geoderma 118:303–320

Liao JH, Wang HH, Tsai CC, Hseu ZY (2006) Litter production, decomposition and nutrient return of uplifted coral reef tropical forest. Forest Ecol Manag 235:174–185

Oze C, Fendorf S, Bird DK, Coleman RG (2004a) Chromium geochemistry of serpentine soils. Intern Geol Rev 46:97–126

Oze C, Fendorf S, Bird DK, Coleman RG (2004b) Chromium geochemistry in serpentinized ultramafic rocks and serpentine soils from the Franciscan complex of California. Am J Sci 304:67–101

Oze C, Skinner C, Schroth A, Coleman RG (2008) Growing up green on serpentine soils: Biogeochemistry of serpentine vegetation in the Central Coast Range of California. Appl Geochem 23:3391–3403

Quevauviller P, Rauret G, Muntau H (1994) Evaluation of a sequential extraction procedure for the determination of extractable trace metal contents in sediments. Fresenius J Anal Chem 349:808–814

4.1 General Information

4.1.1 Geology

The Plio-Plesitocene volcanic activities are the most important in the geologic evolution of Taiwan (Ho 1988). Three prominent volcanic areas are known in Taiwan. The Tatun Volcano Group (TVG) in the north is a group of volcanic cones composed largely of andesitic flows and pyroclastic deposits. These mountains, culminating in Chihshingshan at an elevation of 1,120 m, form the rounded northern extremity of Taiwan. All the volcanoes in this group are dormant and are characterized by numerous fumaroles, solfataras, and hot springs, without any record of active eruption in historic times. The Penghu Islands represent a volcanic area of fissure eruptions. These islands and islets are marked by low flat-topped tablelands consisting of basalt flows and tuffs with minor sand and clay intercalations. The highest elevation in this island group is a little over 50 m above sealevel. The third volcanic area is the Chilung Volcano Group (CVG), consisting of scattered exposures of dacite that form several distinct ridges protruding out from Miocene strata on the northern coast east of Chilung city. Physiographically, little in the way of volcanic landforms can be distinguished. The dacite bodies could be shallow irregular intrusions or extrusions and are closely associated with the gold-copper deposits in that area.

The TVG comprises a series of andesitic volcanoes in northernmost part of Taiwan, the nearest being 15 km north of Taipei. TVG is one of the important Pleistocene andesitic volcano groups, and effusion of andesite lavas commonly alternated with explosive eruption of pyroclastic debris. TVG is composed of typical multivent volcanoes, which consists of at least 20 volcanic cones and domes in an area of 20×20 km (400 km^2) (Song et al. 2000). The volcanic activities can be divided into two periods according to the radiometric dating. It began to erupt in 2.8–2.5 Ma, then ceased about one million years, and resurged in 1.5 Ma and lasted to 0.2 Ma (Song et al. 2000). At least 15 separate lava

flows and three tuff breccias formations have been mapped by Chen and Wu (1971). This group of volcanoes is mainly strato-volcanoes, built by successive eruptions of andesitic flows, volcanic ash, and coarse pyroclastic ejecta. The Tatun andesites commonly contain augite, hornblende hypersthene or combinations of these (Ho 1988).

In 1963, the "Yangmingshan National Park Project" was launched, and the Yangmingshan National Park (YMS-NP) was formally established in 1985. YMS-NP is the only park in Taiwan that has volcanic geography and hot springs. It includes Mountains of Shamao, Chihsing, Tatun, Meintein, Chutzu, and Haungzuei, etc. (Fig. 4.1). Total area of the park is 11,455 ha (Fig. 4.2). Volcanic activity tailed off after the formation of Mt. Shamao about 0.3 million years ago; leaving the various signs of post-volcanic activities we can see today. Volcanic eruptions have caused most of the original sedimentary rock to be covered by volcanic rocks. The volcanic rocks are large andesite of different varieties, with a small amount of basaltic rock.

4.1.2 Climate

According to the report (Chen and Tsai 1983), the important climate factor in the region of YMS-NP included latitude, land and water distribution, prevailing wind, typhoon and tropical depression, baiu rainfall, and topography. Two weather stations located at the region of TVG: the Anpu station locate at the southeastern part of Tatun Mt. with elevation 836 m above sealevel (a.s.l.), and Chutzuhu station locate at the southwestern part of Chishing Mt. with elevation 600 m a.s.l. In the last 30 years (1981–2010), the mean annual air temperature was 16.9 and 18.6 °C in Anpu and Chutzuhu (Fig. 4.3) weather station, respectively. In both stations, the coldest month was January, and the hottest month was July. The mean annual rainfall was about 4,900 and 4,400 mm in Anpu and Chutzuhu weather station, respectively. The rainfall was highest in September (>700 mm), and most of the rainfall was concentrated from

Z.-S. Chen et al., *The Soils of Taiwan*, World Soils Book Series,
DOI 10.1007/978-94-017-9726-9_4, © Springer Science+Business Media Dordrecht 2015

Fig. 4.1 The landscape view of major volvanic mountains in the Yangmingshan national park (YMS-NP). **a** Chihsing Mt.; **b** Tatun Mt.; **c** Huangzuei Mt.; **d** Shamao Mt.; **e** Chutzu Mt.; **f** Meintein Mt. Copyright by Prof. Tsai

September to November (autumn). Typhoon and northeast monsoon both caused very high rainfall during July and December, or till March in the next year. The amount of evaporation in this region was smaller than the rainfall, and annually only 859 mm in Anpu and 951 mm in Chutzuhu station (Chen and Tsai 1983). The mean annual humidity (%) was higher than 85 %, indicating the udic to perudic soil moisture regime in YMS-NP region. Hence, according to the climate data in Anpu and Chutzuhu station, the soil moisture and temperature regimes are perudic and thermic, respectively (Soil Survey Staff 2010). The typhoon and the northeast monsoon have great effects on the soil genesis of this region.

4.1.3 Vegetation and Landuse

The YMS-NP is located in the north of Taiwan, where the northeast monsoon invades perennial. Some habitats are affected by the post-volcanic activity; therefore, there are some distinctive volcanic ecosystems in this park. The biodiversity in this area is high due to the diversity of microenvironment formed by the topography and the direction of the hillside field. There are grassland, temperate deciduous forest, subtropical rain forest, and freshwater ecosystems in this national park. There are more than 1,200 species of vascular plants, occupying about one-third of Taiwan species (Chang and Huang 2001). Dominant herb vegetation types

are broadleaf elatostema (*Elatostema platyphylloides* Shih and Yang), large lineolate elatostema (*Elatostema lineolatum* Forst. var. *majus* Thwait.), Chinese silver grass (*Miscanthus sinensis* Anders. var. *formosanus* Hack.), and silver grass (*Miscanthus floridulus* (Labill.) Warb.). Dominant woody vegetation types are kunishi cane (*Sinobambusa kunishii* (Hayata) Nakai), narrow-petaled hydrangea (*Hydrangea angustipetala* Hayata), dark spotted cherry (*Prunus phaeosticta* (Hance) Maxim.), small-flowered sweet spire (*Itea parviflora* Hemsl.), mori cleyera (*Clevera japonica* Thunb. var. *morri* (Yamamoto) masamune), large-leaved nanmu (*Persea japonica* Sieb.), common machilus (*Persea thunbergii* (Sieb. & Zucc.) Kostermans), and Wheelstaman (*Trochodendron aralioides* Sied & Zucc.). In general, silver grass (*Miscanthus floridulus*) and kunishi cane (*Sinobambusa kunishii*) are grown and distributed at the higher elevation, and common machilus (*Persea thunbergii*) and ryukyu pine (*Pinus luchuensis*) at lower elevation.

4.1.4 Soil Distribution

Based on six pedons selected at elevations between 140 and 1,090 m above sea level (a.s.l.), representing a climatic gradient from about 22 °C mean annual temperature (MAT) and 2,000 mm mean annual precipitation (MAP) to about 16.5 °C MAT and 5,000 mm MAP, Tsai et al. (2010) found that the high-elevation soils (>900 m a.s.l.) contained

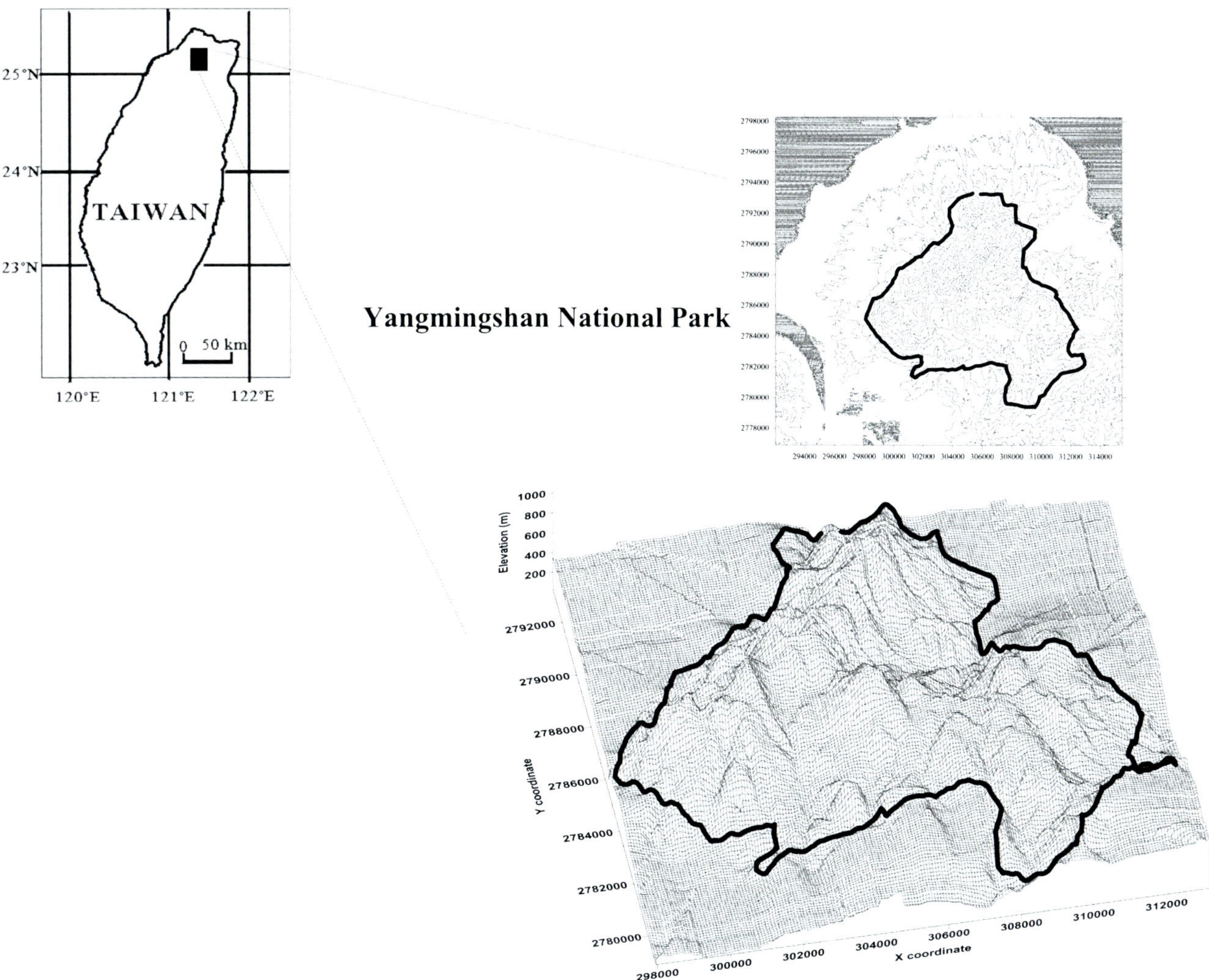

Fig. 4.2 The area of Yangmingshan National Park. Copyright by Prof. Tsai

metastable poorly crystalline materials and classified as Andisols, whereas the low-elevation soils (<300 m a.s.l.) contained thermodynamically more stable minerals and classified as Inceptisols. The soils at intermediate elevation (400–500 m a.s.l.) were at the threshold of andic soil properties and classified as Andic Inceptisols. In addition, according to the results of soil classification of 122 soil pedons collected in the YMS-NP, three Soil Orders (Andisols, Inceptisols, and Ultisols), three Soil Suborders (Udands, Udepts, and Udults), and five Soil Great Groups (Melanudands, Fulvudands, Hapludands, Dystrudepts, and Hapludults) were formed in YMS-NP area (Table 4.1). Among the pedons, 62 pedons were classified as Inceptisols, 57 pedons were Andisols, and only 3 pedons were Ultisols. In general, most of the distribution of Andisols could be found at elevation >700 m a.s.l. in YMS-NP (Fig. 4.4). The distribution of Inceptisols was more dispersed, and could be found at elevation 500–700 m a.s.l. with slope 25–40° or at

elevation 800–1,000 m a.s.l. with slope 5–40°. Ultisols were specific and rare in this volcanic landscape.

4.2   Profile Characteristics

4.2.1   Profile Description

According to Table 4.1, Melanudands, Fulvudands, Hapludands, and Dystrudepts including Andic and Typic subgroups are described in Table 4.2. The first profile described is an Andisols with melanic epipedon from Shamao Mt. It is characteristic in YMS-NP, especially in Shamao Mt. and in some region of Chihsing Mt. The second profile, from Meintein Mt., represents the strong acidic Andisols with very high Al saturation. The third profile, from Tatun Mt., is the common, andesitic Andisols in YMS-NP. The four and five profiles, from meddle to low elevation of Tatun Mt.,

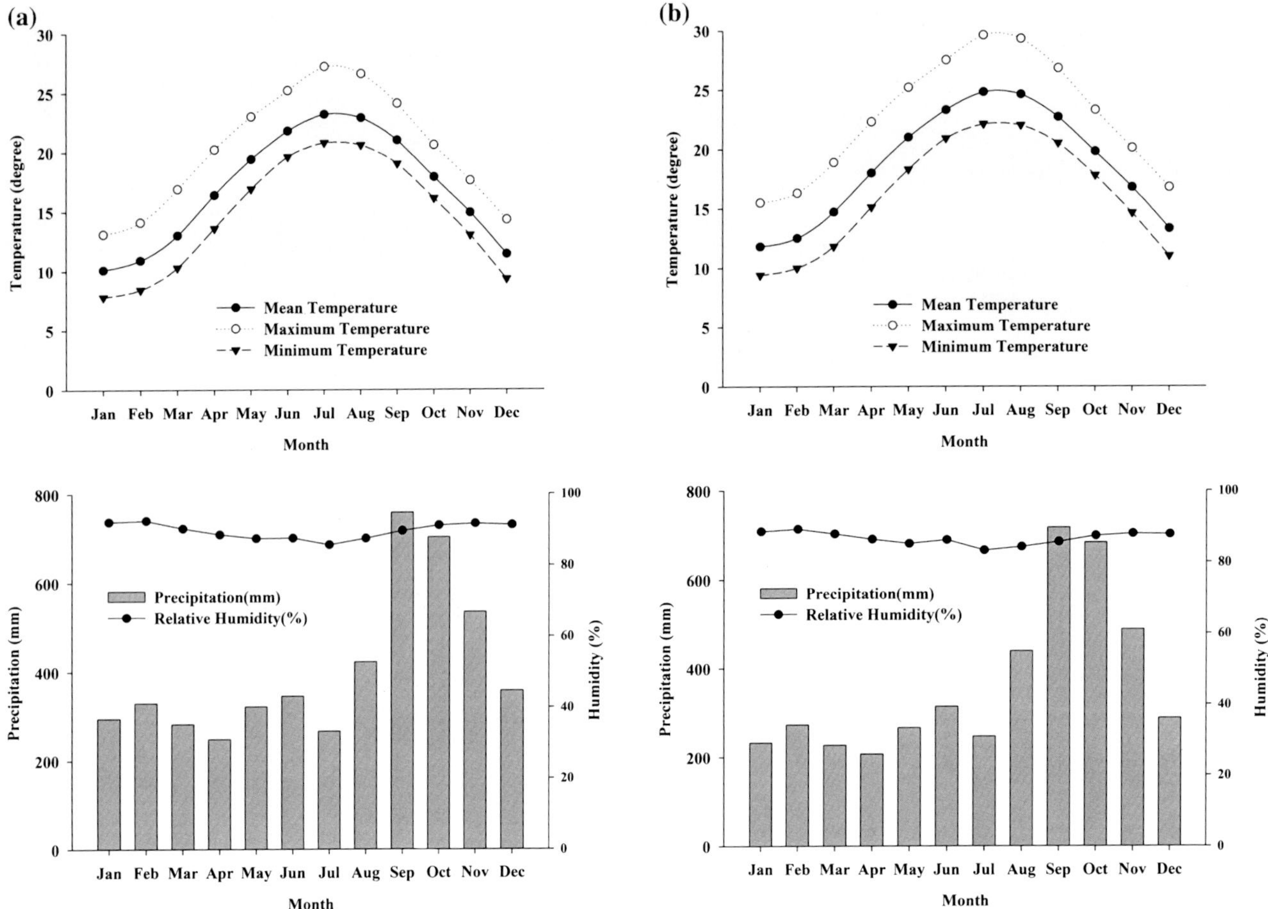

Fig. 4.3 Climate data (1981–2010) of **a** Anpu and **b** Chutzuhu weather station located at Yangmingshan National Park (*Sources* central weather bureau of Taiwan, http://www.cwb.gov.tw/V7/index_home.htm). Copyright by Prof. Tsai

Table 4.1 Soil classification of surveyed 122 soil pedons in YMS-NP

Soil order	No.	Soil suborder	No.	Soil great group	No.
Andisols	57	Udands	57	Melanudands	1
				Fulvudands	5
				Hapludands	51
Inceptisols	62	Udepts	62	Dystrudepts	62
Ultisols	3	Udults	3	Hapludults	3

Copyright by Prof. Tsai

represent the transition from Andisols to Inceptisols. The description of Ultisols is ignored because these soils are uncommon in YMS-NP.

4.2.1.1 Melanudands

The profile described below (Melanudands) is situated in Shamao Mt. and at an elevation of 470 m above sea level (Fig. 4.5a). The mean annual air temperature and annual precipitation are 16 °C and 4,000 mm, respectively. The soil moisture regime and soil temperature regime are per-udic and thermic, respectively. The dominant parent material is hypersthene hornblende andesite. The native vegetation is dominantly large-leaved nanmu (*Persea japonica* Sieb.), common machilus (*Persea thunbergii* (Sieb. & Zucc.) Kostermans), and sugi (*Cryptomeria japonica* (L.f.) D. Don).

Field morphology of the Melanudands pedon is described as follows. If not otherwise indicated, soil color is given for moist soil.

A1	0–26 cm	Black (2.5Y 2/0) loam; weak very fine and fine granular structure; very friable; many very fine and fine roots; diffuse boundary
A2	26–46 cm	Black (2.5Y 2/0) loam; weak very fine angular blocky structure and weak very fine and fine granular structure; very friable; many very fine and fine roots; abrupt smooth boundary
AC	46–67 cm	Dark brown (7.5YR 3/2) loam; weak very fine and fine granular structure; very friable; few very fine and fine roots; clear smooth boundary
C	67–120 cm	Brown (10YR 4/3) loam; weak very fine and fine granular structure; very friable; few very fine and fine roots

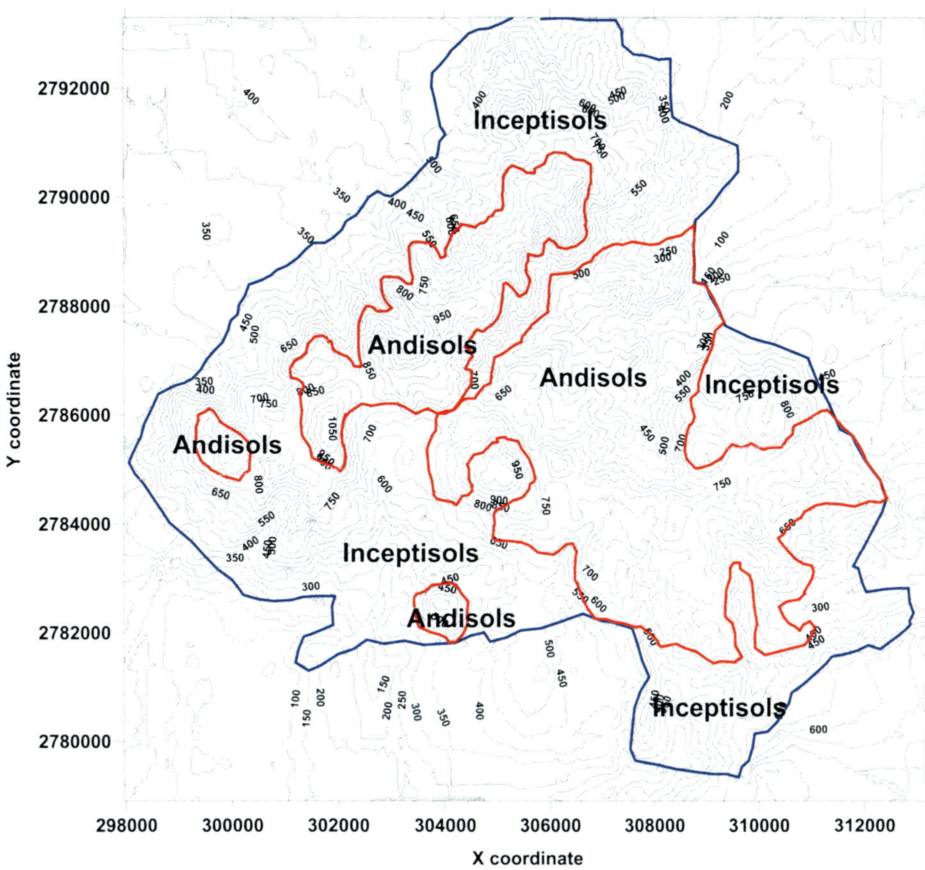

Fig. 4.4 The schematic diagram of distribution of Andisols and Inceptisols in YMS-NP. Copyright by Prof. Tsai

Table 4.2 Site characteristics of the five representative soils

Soil	Elevation (m a.s.l.[a])	Slope (deg.[b])	Aspect (deg.[b])	Parent material	Vegetation type
Melanudands	470	26	–	Hypersthene hornblende andesite	Secondary forest
Fulvudands	800	21	90	Two-pyroxene hornblende andesite	Secondary forest
Hapludands	915	1	135	Hypersthene hornblende andesite	Silver grass
Andic Dystrudepts	440	18	330	Two-pyroxene hornblende andesite	Secondary forest
Typic Dystrudepts	140	30	180	Andesitic lower tuff breccia	Secondary forest

[a] *a.s.l.* above sea level
[b] *deg.* degrees
Copyright by Prof. Tsai

4.2.1.2 Fulvudands

The profile described below (Fulvudands) is situated in Meintein Mt. and at an elevation of 800 m above sea level (Fig. 4.5b). The area of Meintein Mt. has an elevation ranging from 800 to about 1,000 m and a slope varying from 10 to more than 30 % (Chen et al. 1999). The geology is characterized by consolidated andesitic pyroclastic rocks rich in amphibole and pyroxene erupted about 300,000–800,000 years B.P. (Chen 1990). Geologic records of the area (Chen 1990) and actual field observations indicate that the volcanic ash blanketing the study area is of almost the same geochemistry and several events. Climatic seasonal records for the mountain reveal an average annual rainfall of 4,800 mm, average temperature of 16.6 °C (min. = 14.3 °C, max. = 19.7 °C), and average relative humidity of 92 %. The moisture and temperature regimes are perudic and thermic, respectively. The natural vegetation cover is dominantly silver grass (*Miscanthus floridulus*) and kunishi cane (*Sinobambusa kunishii*) in the upper slopes and common machilus (*Persea thunbergii*) and ryukyu pine (*Pinus luchuensis*) in the lower slopes. Some local species of bracken fern also grow throughout the area.

Fig. 4.5 Soil morphology of five representative soils in YMS-NP. **a** Melanudands; **b** Fulvudands; **c** Hapludands; **d** Anidc Dystrudepts; **e** Typic Dystrudepts. Copyright by Prof. Tsai

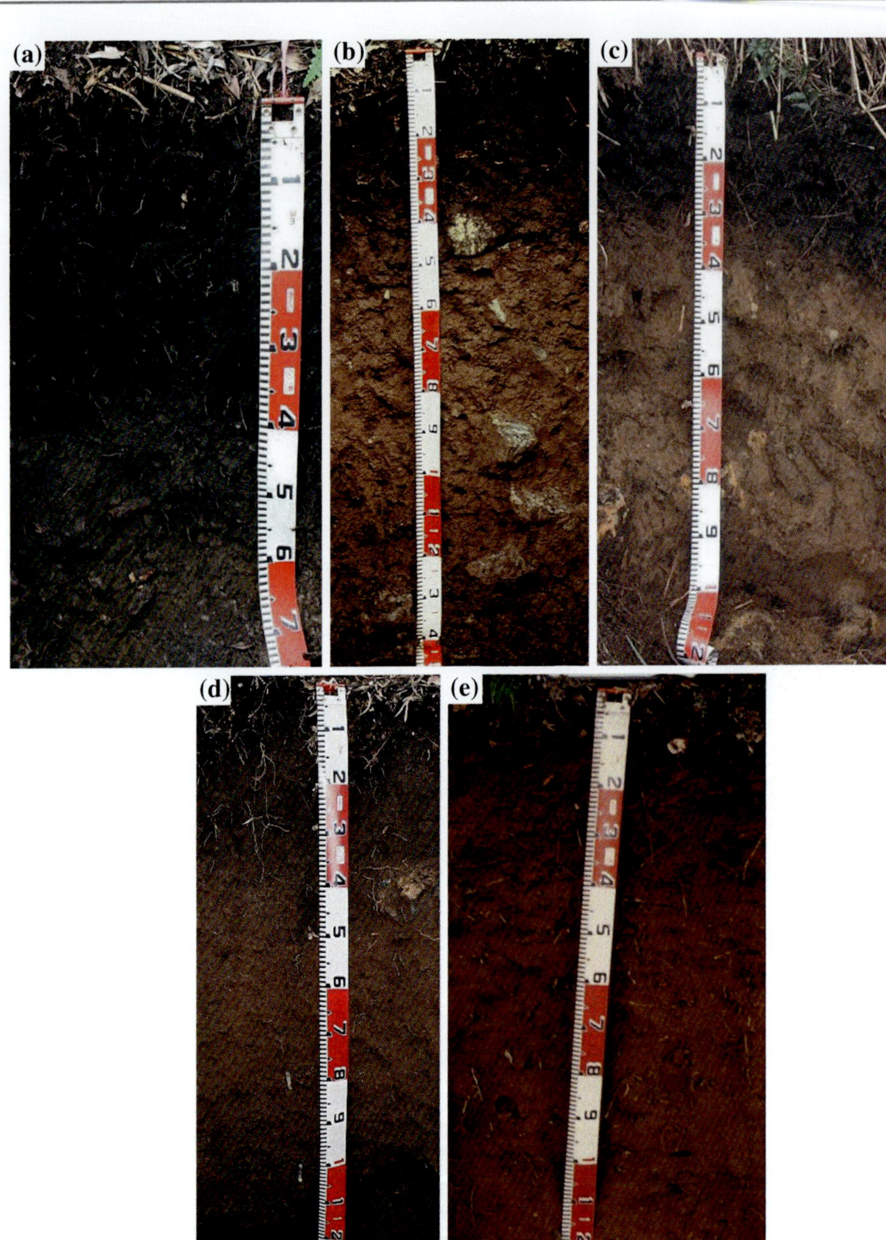

Field morphology of the Fulvudands pedon is described as follows. If not otherwise indicated, soil color is given for moist soil.

O/A1	0–15 cm	Black (5YR 2.5/1) loam; moderate very fine and fine granular structure; very friable; many very fine and fine roots; diffuse boundary
A2	15–30 cm	Reddish black (2.5YR 2.5/1) loam; moderate very fine and fine granular structure, and moderate very fine and fine angular blocky structure; firm; many very fine and fine roots; gradual boundary
Bw1	30–48 cm	Dark reddish brown (5YR 3/3) clay loam; moderate fine granular structure, and moderate fine and medium angular blocky structure; firm, slightly sticky and slightly plastic; some very fine and fine roots, and few medium roots; abrupt smooth boundary
Bw2	48–83 cm	Reddish brown (5YR 4/4) clay loam; moderate very fine, fine and medium angular blocky structure; firm, slightly sticky and slightly plastic; some very fine and fine roots, and few medium roots; diffuse boundary
Bw3	83–110 cm	Reddish brown (5YR 4/4) clay loam; moderate very fine, fine and medium angular blocky structure; firm, slightly sticky and slightly plastic; few very fine roots, and few medium roots; diffuse boundary
BC	110–140 cm	Reddish brown (5YR 4/4) clay loam; moderate very fine, fine and medium angular blocky structure; firm, sticky and plastic; few very fine and fine roots; clear smooth boundary
C	>140 cm	Reddish brown (5YR 4/4) clay; moderate very fine and fine angular blocky structure; firm, sticky and plastic

4.2.1.3 Hapludands

The profile described below (Hapludands) is situated in Tatun Mt. and at an elevation of 915 m above sea level (Fig. 4.5c). The mean annual air temperature and annual precipitation at Anpu station are 16.7 °C (monthly min. = 9.8 °C, monthly max. = 23.2 °C) and 4,800 mm, respectively. According to Chen and Tsai (1983), the mean annual potential evapotranspiration is about 18 % of the mean annual rainfall at Anpu station (data recorded from 1962 to 1981), which resulted in a mean annual effective precipitation of approximately 4,000 mm at Anpu. The soil moisture regime of the study area is udic to perudic, and the soil temperature regime is thermic to hyperthermic (Soil Survey Staff 2010). The dominant vegetation types are silver grass (*Miscanthus floridulus*) and kunishi cane (*Sinobambusa kunishii*).

Field morphology of the Hapludands pedon is described as follows. If not otherwise indicated, soil color is given for moist soil.

A1	0–13 cm	Black (7.5YR 2/0) loam; weak very fine and fine subangular blocky structure, and weak very fine and fine granular structure; loose to very friable; many very fine and fine roots; diffuse boundary
A2	13–30 cm	Black (7.5YR 2/0) silt loam; weak very fine and fine subangular blocky structure, and weak very fine and fine granular structure; loose to very friable; many very fine and fine roots; abrupt smooth boundary
Bw1	30–47 cm	Reddish brown (5YR 4/4) loam; weak very fine and fine granular structure, and moderate very fine and fine subangular blocky structure; firm, many very fine roots; gradual boundary
Bw2	47–65 cm	Strong brown (7.5YR 5/8) loam; weak very fine and fine granular structure, and moderate very fine and fine subangular blocky structure; firm, few very fine roots; clear smooth boundary
C	>65 cm	Yellowish brown (10YR 5/8) loam; weak very fine and fine granular structure, and moderate very fine and fine subangular blocky structure; firm, few very fine roots

4.2.1.4 Andic Dystrudepts

The profile described below (Andic Dystrudepts) is situated in the lower backslope of Tatun Mt. and at an elevation of 440 m above sea level (Fig. 4.5d). Because no weather station existed in this region, the climatic data are estimated from two meteorological stations, located at 19 m a.s.l. (Danshui station) and 836 m a.s.l. (Anpu station), respectively. Mean annual rainfall (record: 1971–2000) ranges from 2,120 mm at Danshui station to 4,892 mm at Anpu station, and mean annual temperature from 22.1 °C (monthly min. = 15.1 °C, monthly max. = 28.8 °C) at Danshui station to 16.7 °C (monthly min. = 9.8 °C, monthly max. = 23.2 °C) at Anpu station. The soil moisture regime is udic, and soil temperature regime is hyperthermic (Soil Survey Staff 2010). The dominant vegetation types are common machilus (*Persea thunbergii*) and ryukyu pine (*Pinus luchuensis*).

Field morphology of the Andic Dystrudepts pedon is described as follows. If not otherwise indicated, soil color is given for moist soil.

A	0–10 cm	Very dark brown (7.5YR 2.5/3) loam; moderate very fine granular structure; very friable; many very fine roots, and common medium roots; gradual smooth boundary
AB	10–30 cm	Dark brown (7.5YR 3/4) loam; moderate very fine and fine granular structure; friable; many very fine and fine roots; diffuse boundary
BA	30–40 cm	Dark brown (7.5YR 3/4) clay loam; moderate very fine granular structure, and moderate very fine angular blocky structure; firm, slightly sticky and slightly plastic; common fine roots; diffuse boundary
Bw1	40–65 cm	Dark brown (7.5YR 3/4) clay loam; moderate very fine and fine angular blocky structure; firm, slightly sticky and slightly plastic; common very fine and fine roots; diffuse boundary
Bw2	65–100 cm	Dark brown (7.5YR 3/4) clay loam; moderate very fine and fine angular blocky structure; firm, sticky and plastic; few very fine roots; diffuse boundary
Bw3	100–125 cm	Dark brown (7.5YR 3/3) clay loam; moderate very fine and fine angular blocky structure; very firm, sticky and plastic; few very fine and fine roots; diffuse boundary
BC	125–150 cm	Dark brown (7.5YR 3/4) clay loam; moderate very fine angular blocky structure; firm, sticky and plastic; diffuse boundary
C	>150 cm	Dark yellowish brown (10YR 4/4) clay; moderate very fine and fine angular blocky structure; firm, sticky and plastic

4.2.1.5 Typic Dystrudepts

The profile described below (Typic Dystrudepts) is also situated in the lower backslope of Tatun Mt. and at an elevation of 140 m above sea level (Fig. 4.5e). The climate, soil moisture regime, soil temperature regime, and dominant vegetation types are similar to above (Andic Dystrudepts). Field morphology of the Typic Dystrudepts pedon is described as follows. If not otherwise indicated, soil color is given for moist soil.

A	0–10 cm	Dark brown (7.5YR 3/3) loam; moderate very fine granular structure; no sticky and no plastic; many very fine, fine and medium roots, and common very coarse roots; clear smooth boundary
Bw1	10–23 cm	Brown (7.5YR 4/4) loam; weak very fine granular structure; loose to very friable, no sticky and no plastic; many very fine, fine, and medium roots, and few coarse roots; diffuse boundary
Bw2	23–50 cm	Strong brown (7.5YR 4/6) silt loam; weak very fine and fine granular structure; loose, slightly sticky and slightly plastic; many very fine and fine roots, and common medium roots; diffuse boundary
Bw3	50–70 cm	Strong brown (7.5YR 4/6) slit loam; weak very fine angular blocky structure, and weak very fine granular structure; very friable, slightly sticky and slightly plastic; many very fine, fine, and medium roots; diffuse boundary
Bw4	70–90 cm	Strong brown (7.5YR 4/6) loam to silt loam; weak very fine and fine angular blocky structure; friable, slightly sticky and slightly plastic; many very fine roots, and common fine and medium roots; diffuse boundary
Bw5	90–115 cm	Strong brown (7.5YR 4/6) loam to silt loam; weak very fine and fine angular blocky structure; friable, slightly sticky and slightly plastic; common very fine and fine roots; gradual smooth boundary
C	115–140 cm	Strong brown (7.5YR 5/6) loam

4.2.2 Characteristics of Pyroclastic-Derived Soils

4.2.2.1 Morphology

Some unique physical properties of Andisols are directly visible to the eye and sensible to the touch, i.e., surface soils are rich in humus and dark in color, soil clods are light, fluffy, and easy to break into small pieces (Nanzyo 2003). Melanudands from Shamao Mt. has characteristic melanic epipedon with thickness more than 45 cm; soil has an A-AC-C horizon sequence; A horizon is the black Munsell color (2.5Y 2/0); soils are moderately deep; texture is loamy and the structure is granular in all horizons; fine and very fine roots are abundant in the upper horizons. Frequently they are stony or gravelly, depending on slope. Fulvudands has an A-Bw-BC-C horizon sequence with a solum thickness 140 cm. The soils had a color hue of 2.5YR–5YR, and a texture ranging from loam in the A horizons to clay loam in the Bw horizons. The soils are medium-textured as a result of being loam in surface and clay loam in subsurface horizons. Structure is granular in the surface grading to moderate subangular blocky in the subsoil. Consistency was very friable in surface and firm with a slight tendency to be plastic and sticky in subsurface horizons. Very fine roots were abundant on the surface and could still be observed even in BC horizons, but not in C horizons.

For Hapludands, Andic Dystrudepts, and Typic Dystrudepts, all soils had thick (10–30 cm) black (7.5YR 2/0) to dark brown (7.5YR 3/3) granular A horizons with very friable moist consistence and were highly permeable, which is a first indication that andic soil properties might still be present. In addition, as suggested by Tsai et al. (2010), with decreasing elevation, surface soil color changed from black to dark brown or brown. The Bw horizons had subangular blocky structure only in the high-elevation pedons, but dominantly angular blocky structure in the lower-elevation. The shift of vegetation type from silver grass at higher elevations to secondary forest at intermediate and low elevations may be the dominant factor in altering soil color of A horizons and soil structure of Bw horizons.

4.2.2.2 Physicochemical Characteristics

Melanudands, with lower pH (4.1–4.5) and higher Al_p/Al_o ratios (0.7–1.0), was classified as the nonallophanic Andisols. The characteristics are listed as (Table 4.3): thickness of melanic epipedon is more than 45 cm; black Munsell color (2.5Y 2/0); low bulk density (0.37–0.43 Mg m^{-3}); very friable consistency; high organic carbon (10.9–15.6 %); high exchangeable aluminum (3.8–8.0 cmol(+) kg^{-1} soils); high phosphate retention (>99 %); low melanic index (<1.65); high oxalate extractable Al and Fe (Al_o + 1/2 Fe_o = 2.58–3.55 %);

aluminum was mostly complexed with organic matter; only about 5 % allophane in the melanic epipedon; and clay mineral of the parent materials was predominantly imogolite-like.

Results of physical analysis indicate that Fulvudands are dominated by the silt fraction (Table 4.4). Clay was much less than silt and fine sand fractions exceeded coarse sand fractions. Bulk density values were lower than 0.9 Mg/m^3 in most horizons, with the lowest values in surface horizons. Moisture retained at 33 and 1,500 kPa tensions is higher in surface than in subsurface horizons. In addition, the soils are acidic, with pH (H_2O) ranging from 4.1 to 5.0, and pH (KCl) from 3.6 to 4.1. The pH (NaF) was close to or higher than 9.0. Organic C was highest in the surface horizon and decreased with depth in all soils. Exchangeable bases were low, particularly in the surface horizons, resulting in a low base saturation of generally <15 %. In contrast, exchangeable Al in the soils was high (1.4–3.4 cmol(+) kg^{-1} soil), resulting in very high Al saturation ranging from 33 to 76 %. Phosphate retention ranged from 88 to 96 %. The results of selective dissolution analyses show that Fe_o (oxalate extractable Fe) as well as Al_o (oxalate extractable Al) are not vary considerably among soils. Fe_p (pyrophosphate extractable Fe) is generally comparable to Fe_o in most horizons. Al_p (pyrophosphate extractable Al) is higher in surface than in subsurface horizons. In most upper horizons of all soils, the ratios of Fe_p to Fe_o ($Fe_{p/o}$) and of Al_p to Al_o ($Al_{p/o}$), are close to 1.0. Si_o (oxalate extractable Si) was much lower than Fe_o and Al_o. The value of (Al_o + 1/2Fe_o) varied between 2.6 and 3.3 %. Amounts of allophane estimated from selective dissolution data, and showed that allophane were little noted in the A horizon. However, they were present in substantial amounts (1.9–3.4 % g/kg) in the subsurface horizons.

Hapludands, Andic Dystrudepts, and Typic Dystrudepts are clay-rich in accordance with their relatively long exposure to the agents of weathering (Table 4.5). All pedons are clayey throughout, except for the sandy loam C horizon in Hapludands, but no significant clay accumulation can be

Table 4.3 Some physicochemical characteristics of Melanudands

Horizon	Depth (cm)	pH		NaF	O.C (%)	Exch.		BS (%)	Phosphate retention (%)
		H_2O	KCl			CEC_7 (cmol(+) kg^{-1} soil)	Al (cmol(+) kg^{-1} soil)		
A1	0–26	4.1	3.6	9.1	15.6	53.87	8.06	2	99
A2	26–46	4.1	3.9	11.0	10.9	43.30	3.79	1	99
AC	46–67	4.5	4.4	11.2	4.6	37.43	1.48	1	–
C	67–120	4.5	4.6	11.1	2.9	30.18	0.89	1	–

Horizon	Depth (cm)	Al_p (%)	Al_o (%)	Al_p/Al_o	Fe_p (%)	Fe_o (%)	Fe_p/Fe_o	Si_o (%)	Al_o + 1/2Fe_o (%)	Allophane (%)
A1	0–26	1.72	1.61	1.0	1.06	1.34	0.8	0.18	2.58	1
A2	26–46	1.92	2.82	0.7	1.16	1.47	0.8	0.60	3.55	5
AC	46–67	1.01	6.52	0.2	0.18	0.85	0.2	2.69	6.94	22
C	67–120	0.73	7.08	0.1	0.05	0.43	0.1	3.40	7.30	27

Abstracted from Chen and Huang (1991)

Table 4.4 Some physicochemical characteristics of Fulvudands

Horizon	Depth (cm)	Sand fraction					Sand (g kg^{-1})	Silt (g kg^{-1})	Clay (g kg^{-1})	Bd (Mg m^{-3})	Water retention		15 bar/ clay
		VCS (g kg^{-1})	CS (g kg^{-1})	MS (g kg^{-1})	FS (g kg^{-1})	VFS (g kg^{-1})					1/3 bar (%)	15 bar (%)	
O/A1	0–15	3	7	14	29	41	93	486	421	0.48	45	35	0.8
A2	15–30	2	8	16	33	54	112	507	381	0.71	45	33	0.9
Bw1	30–48	5	10	18	64	319	411	338	251	0.78	34	26	1.0
Bw2	48–83	7	9	24	120	172	326	345	329	0.94	33	25	0.8
Bw3	83–110	6	11	62	300	146	515	210	275	–	34	25	0.9
BC	110–140	5	8	14	68	351	437	255	308	–	33	25	0.8
C	>140	5	7	14	119	211	349	292	359	–	35	26	0.7

Horizon	Depth (cm)	pH			O.C (%)	Exchangeable bases				CEC$_7$ (cmol(+) kg^{-1} soil)	Exch. Al (cmol(+) kg^{-1} soil)	BS (%)	Al Sat. (%)	Phosphate retention (%)
		H$_2$O	KCl	NaF		Ca (cmol(+) kg^{-1} soil)	Mg (cmol(+) kg^{-1} soil)	K (cmol(+) kg^{-1} soil)	Na (cmol(+) kg^{-1} soil)					
O/A1	0–15	4.1	3.6	8.3	7.57	1.18	0.77	0.37	0.13	36.35	3.02	7	55	88
A2	15–30	4.2	3.6	9.3	5.98	0.44	0.31	0.19	0.15	31.95	3.41	3	76	90
Bw1	30–48	4.3	3.8	9.6	3.18	0.36	0.29	0.12	0.24	27.46	2.11	4	68	94
Bw2	48–83	4.7	3.9	9.4	1.76	0.94	0.82	0.18	0.16	24.04	1.35	9	39	93
Bw3	83–110	4.8	4.0	9.3	1.55	0.81	1.58	0.12	0.19	24.78	1.86	11	41	96
BC	110–140	5.0	4.1	9.3	1.42	0.78	1.84	0.13	0.22	24.88	1.48	12	33	95
C	>140	5.0	4.0	9.4	1.31	0.58	1.65	0.22	0.23	21.24	2.11	13	44	94

Horizon	Depth (cm)	Al$_p$ (%)	Al$_o$ (%)	Al$_p$/Al$_o$	Fe$_p$ (%)	Fe$_o$ (%)	Fe$_p$/Fe$_o$	Si$_o$ (%)	Al$_o$ + 1/2Fe$_o$ (%)	Allophane (%)
O/A1	0–15	1.37	1.59	0.86	2.27	2.00	1.13	0.19	2.59	0.9
A2	15–30	1.43	1.69	0.85	2.36	2.15	1.10	0.21	2.76	1.0
Bw1	30–48	1.19	1.92	0.62	2.56	2.31	1.11	0.24	3.08	2.9
Bw2	48–83	1.13	1.62	0.70	2.19	2.18	1.00	0.19	2.71	1.9
Bw3	83–110	1.02	2.03	0.50	2.16	2.48	0.87	0.21	3.27	3.4
BC	110–140	1.13	2.03	0.56	1.93	2.43	0.79	0.21	3.24	3.4
C	>140	0.96	1.55	0.62	1.77	2.13	0.83	0.20	2.61	2.4

Abstracted from Chen et al. (1999)

Table 4.5 Selected physical and chemical characteristics of Hapludands, Andic Dystrudepts, and Typic Dystrudepts

Horizon	Sand (%)	Clay (%)	Est. clay[a] (%)	Bd (Mg m^{-3})	H$_2$O retention		pH		OC (%)	ECEC (cmol$_c$ kg^{-1})	Al sat. (%)	P ret. (%)
					33 kPa (%)	1,500 kPa (%)	H$_2$O (%)	NaF (%)				
Hapludands												
A1	17	51	61	0.27	45	38	4.5	9.2	13	22	78	82
A2	14	55	62	0.38	40	35	4.5	9.4	11	19	89	88
Bw1	25	55	42	0.59	33	24	4.8	10.8	3.8	3.7	70	94
Bw2	34	41	62	0.72	41	27	4.7	10.8	2.4	3.8	64	93
C	58	12	60	0.84	44	26	4.7	10.6	0.9	2.0	31	96
Andic Dystrudepts												
A	8.3	70	61	0.66	42	33	3.8	9.7	7.7	21	69	75
AB	7.5	69	47	0.79	43	27	4.2	9.8	3.5	16	74	85
BA	6.3	75	56	0.77	40	26	4.2	10.9	2.7	8.3	43	88
Bw1	6.5	69	55	0.87	30	25	4.3	10.7	1.9	9.2	31	87
Bw2	6.2	73	56	1.02	31	25	5.1	10.6	1.2	6.0	13	85
Bw3	6.4	64	59	1.14	31	25	5.5	10.7	1.3	6.5	8	85
BC	6.4	69	62	1.48	33	27	5.6	10.8	1.5	6.7	9	87
C	7.7	63	62	–	32	26	–	–	1.6	7.4	9	87
Typic Dystrudepts												
A	18	63	52	0.58	35	27	3.9	8.0	6.8	9.8	66	12
Bw1	12	74	51	0.82	35	24	4.2	8.6	2.4	6.7	79	24
Bw2	12	67	63	0.84	36	28	4.4	9.7	2.2	6.6	84	71
Bw3	14	67	60	0.81	37	27	4.5	9.7	2.1	6.4	88	68
Bw4	16	65	57	0.82	36	25	4.5	9.9	1.9	5.1	80	66
Bw5	19	62	63	0.91	37	28	4.6	10.2	1.8	4.0	84	78
C	35	37	64	0.81	42	27	4.7	10.6	0.9	4.3	67	94

Est. clay estimated clay, *Bd* bulk density, *OC* organic carbon, *ECEC* effective cation exchange capacity, *Al sat.* Al saturation, *P ret.* phosphate retention, – not determined

[a] Estimated clay (%) = (% H$_2$O retained at 1,500 kPa − % organic C) × 2.5 (Soil Survey Staff 2006)

Abstracted from Tsai et al. (2010)

found. Kao (2003) have revealed no evidence of clay films and thus suggest that, despite the high leaching environment, clay was formed and weathered in situ. Estimated clay, based on Soil Survey Staff (2006), resulted in similarly high clay contents and thus confirmed the measured data. All horizons of Hapludands have oven-dry bulk density (Bd) of <0.85 Mg m^{-3}. The Bd of Andic Dystrudepts is clearly higher, reaching values >1 Mg m^{-3}. However, the Bd of Typic Dystrudepts is again relatively low, i.e. generally <0.85 Mg m^{-3}. Water retention at 33 and 1,500 kPa indicates that Hapludands has higher water holding capacity than Dystrudepts. Water retention at 1,500 kPa is relatively high for Hapludands, and is an indication of the presence of allophane and allophane-like materials with small particle-sizes and hollow spherical structures (Shoji et al. 1993). The pH (H$_2$O) values of the three soils are less than 5.0. Shoji and Fujiwara (1984) proposed that at pH values of 5 or less, organic acids are the dominant proton donors in volcanic ash

soils, lowering aqueous Al activities through the formation of Al-humus complexes. The pH (NaF) values of the three soils were mostly above 9.5, which suggest that amorphous materials or active Al-OH groups are dominant in the soil exchange complexes. This indicates that the volcanic parent materials may still influence the soil characteristics, even though the soils have undergone long-term and intense weathering in this study area (Tsai et al. 2010).

The OC contents of surface soil was >10 % in Hapludands. The topsoils of Dystrudepts had significantly lower OC contents (2.7–7.7 %). High additions of organic matter from the silver grass vegetation combined with slowed decomposition due to lower temperatures, low pH values, and the presence of stabilizing Al-humus complexes may have caused the increased accumulation of organic C at Hapludands. Conversely, the lower organic matter contents at Dystrudepts were presumably the result of less addition in secondary forest combined with more rapid decomposition

Table 4.6 Results of selective dissolution analysis and calculated allophane and ferrihydrite contents in Hapludands, Andic Dystrudepts, and Typic Dystrudepts

Horizon	Fe_d (g kg^{-1})	Fe_o (g kg^{-1})	Al_o (g kg^{-1})	Si_o (g kg^{-1})	Fe_o/Fe_d	Fe_p/Fe_o	Al_p/Al_o	$(Al_o - Al_p)/Si_o$	Allo. (%)	$Al_o + 1/2Fe_o$ (%)	Ferr. (%)
Hapludands											
A1	20.8	20.0	15.6	9.3	0.96	0.19	0.53	0.78	4.79	2.6	3.4
A2	27.3	20.3	15.1	9.0	0.74	0.19	0.60	0.67	4.50	2.5	3.5
Bw1	30.1	23.5	30.8	17.6	0.78	0.13	0.23	1.34	10.6	4.3	4.0
Bw2	24.7	23.3	25.2	14.5	0.94	0.07	0.18	1.42	8.98	3.7	4.0
C	15.8	21.6	19.7	11.5	1.37	0.03	0.14	1.48	7.25	3.1	3.7
Andic Dystrudepts											
A	43.3	19.7	13.5	3.3	0.45	0.58	0.92	0.33	1.52	2.3	3.3
AB	51.0	19.2	15.7	4.2	0.38	0.48	0.73	1.02	2.31	2.5	3.3
BA	46.0	20.2	15.7	4.0	0.44	0.47	0.68	1.25	2.35	2.6	3.4
Bw1	47.3	17.8	13.1	3.0	0.38	0.74	0.81	0.83	1.57	2.2	3.0
Bw2	52.2	17.5	12.8	2.8	0.34	0.53	0.59	1.89	2.04	2.2	3.0
Bw3	52.0	19.4	12.9	3.5	0.37	0.45	0.54	1.69	2.36	2.3	3.3
BC	51.7	19.1	13.6	4.1	0.37	0.40	0.49	1.71	2.79	2.3	3.2
C	56.3	17.6	15.0	5.2	0.31	0.38	0.43	1.65	3.47	2.4	3.0
Typic Dystrudepts											
A	40.2	5.0	2.3	1.6	0.12	0.62	0.96	0.06	–	0.5	0.9
Bw1	46.5	4.9	3.1	2.0	0.11	0.69	0.77	0.35	–	0.6	0.8
Bw2	50.4	5.0	5.6	4.1	0.10	0.90	0.61	0.54	–	0.8	0.9
Bw3	42.6	5.0	5.6	5.6	0.12	0.90	0.63	0.38	–	0.8	0.9
Bw4	45.9	4.2	5.5	6.0	0.09	1.12	0.65	0.32	–	0.8	0.7
Bw5	45.9	4.1	6.3	5.9	0.09	0.88	0.52	0.51	–	0.8	0.7
C	43.9	2.7	6.2	5.4	0.06	0.33	0.66	0.39	–	0.8	0.5

Fe$_d$ dithionite-extractable Fe; *Fe$_o$, Al$_o$, Si$_o$* acid-oxalate-extractable Fe, Al, and Si; *Fe$_p$, Al$_p$* sodium pyrophosphate-extractable Fe and Al; *Allo.* Allophane (%) = $100 \times \%Si_o/[23.4 - 5.1x]$ with $x = (Al_o - Al_p)/Si_o$ (Parfitt and Wilson 1985; Mizota and van Reeuwijk 1989); *Ferr* Ferrihydrite (%) = $1.7 \% \times Fe_o$ (Childs et al. 1991); – not determined
Abstracted from Tsai et al. (2010)

due to higher temperatures and lower amounts of stabilizing Al-humus complexes. Effective cation exchange capacity (ECEC) showed a similar pattern with OC. Hapludands and Typic Dystrudepts have the highest Al saturation, mostly >60 %. The highest phosphate retention was found in Hapludands. High phosphate retention indicates the presence of amorphous materials.

The results of the selective dissolution analyses (Table 4.6) show that the values for oxalate-extractable Fe, Al, and Si (Fe_o, Al_o and Si_o) were strikingly high in Hapludands, highlighting the abundance of poorly crystalline and non-crystalline materials. In Andic Dystrudepts, the values of Fe_o and Al_o were lower than in Hapludands but higher than in Typic Dystrudepts. Generally, progressing weathering and soil formation lead to a decrease in Fe_o/Fe_d ratio with soil age (Lair et al. 2009). In these three soils, the Fe_o/Fe_d ratios decreased with decreasing elevation, suggesting that amorphous Fe predominates at high elevation (Hapludands). There, the higher OC concentrations and

lower temperatures may inhibit Fe oxide crystallization (Schwertmann 1985). The Fe oxide contents (Fe_d) notably increased in Andic and Typic Dystrudepts, and we speculate that the warmer climate has promoted primary mineral weathering and the formation of Fe oxides. The ratio Fe_p/Fe_o was ≤0.20 in Hapludands, but increased with decreasing elevation. In Hapludands, the Al_p/Al_o ratio was >0.50 in the A1 and A2 horizons and ranged from 0.14 to 0.23 in the subsoil horizons. This suggests that Al was dominantly in humus complexes in the topsoil, but incorporated mainly into short-range-order alumosilicates (e.g. allophane and imogolite) in the subsoil (Vacca et al. 2003). In Andic and Typic Dystrudepts, the Al_p/Al_o ratio was generally >0.50, showing that Al-humus complexes were dominant over allophane and imogolite. With some exceptions, the $(Al_o - Al_p)/Si_o$ ratio was between 1 and 2 in Hapludands and Andic Dystrudepts, which is in the range of ratios expected for allophane and imogolite (Shoji et al. 1993), but clearly below this range in Typic Dystrudepts. Allophane contents were highest in Hapludands, where significant amounts were found in the entire profile, ranging from 4 to 11 % and increasing with depth. In Andic Dystrudepts, allophane is found in considerably smaller quantities (<3 %) and predominantly in the lower part of the profile. In Typic Dystrudepts with $(Al_o - Al_p)/Si_o$ ratio mostly less than 0.5, allophane contents were not determined. The percentage of $(Al_o + 1/2Fe_o)$, one of the criteria for Andisols classification, was between 2.5 and 4.3 % in Hapludands, and around 2 % in Andic Dystrudepts, suggesting that active forms of Al and Fe have accumulated in these soils. In Typic Dystrudepts, on the other hand, the percentages were not determined, indicating that active Al and Fe have changed (to more crystalline forms) in the warmer and less moist environment. Ferrihydrite is a poorly crystalline to non-crystalline Fe oxide and has a brownish color in Bw horizons of Andisols (Nanzyo 2003). In these three soils, the estimated contents and distribution of ferrihydrite were similar to the distribution of allophane and $(Al_o + 1/2Fe_o)$.

4.2.2.3 Mineral Composition

In Fulvudands, X-ray diffraction analysis of the sand fraction revealed dominant amounts of feldspar, moderate quantities of amphibole, pyroxene, and quartz, and minor quantities of cristobalite and gibbsite in soil (Table 4.7). The clay fraction revealed an abundance of kaolinite, gibbsite, and quartz. Minor quantities of cristobalite and 2:1 layer silicates (illite) were also noted. All three layer silicates were detected in Fulvudands soil. As indicated by Chen et al. (1999), transmission electron microscopy studies on selected samples confirmed the presence of kaolinite, halloysite, gibbsite, allophane, and imogolite. Kaolinite appeared as flakes of varying sizes, with some showing typical hexagonal shapes, whereas halloysite was tubular, and gibbsite was warped rods and flakes, both morphologies in accordance with Dixon (1989). Allophane assumed the aggregate form, and imogolite was observed as smooth and curved threads similar to the description of Wada (1989). Opaline silica, a common early product of pyroclastic weathering, was not observed although previous studies conducted in the volcanic mountains near the study site did observe it (Huang and Chen 1990, 1992).

The bulk soil XRD data (Table 4.8) confirmed the presence of gibbsite and showed increasing intensities of a 7Å-reflection from Hapludands to Typic Dystrudepts. Similarly, hematite was absent in Hapludands, appeared in Andic Dystrudepts and further increased in Typic Dystrudepts. Conversely, amphibole was only detected in Hapludands, but was absent in Dystrudepts. This again indicates that the warmer climate at lower elevation has promoted primary mineral weathering and the formation of crystalline Fe oxides (Tsai et al. 2010). Quartz and cristobalite were found in all pedons. In addition, semiquantitative estimations of clay mineral abundance (based on <2 μm XRD) are also presented in Table 3.6. We did not find smectite and only traces of vermiculite and mixed layered minerals in the studied pedons. Illite and kaolin minerals tended to increase from Hapludands to Typic Dystrudepts. Conversely, chlorite tended to decrease from Hapludands to Typic Dystrudepts. Significant amounts of gibbsite were detected in all pedons.

4.2.3 Formation and Genesis

In the Yangmingshan Volcanic National Park (YMS-NP), the early studies were conducted for volcanic soils and related soils in the 1970s. But the detailed and systematic investigations were conducted after the 1990s. Including soils developed from different volcanic parent materials, micromorphological approaches in studying andesite weathering, soil-landscape relationships and soil mapping, and transition soils from Andisols to Ultisols along the topographic sequence are important study subjects in YMS-NP.

In the region of Tatun Mt., soils derived from different volcanic parent materials including volcanic detritus and two pyroxene andesite (Chang 1972), two pyroxene-hornblende

Table 4.7 Identification and semi-quantitaton of minerals in sand and clay fraction by XRD

Horizon	Depth (cm)	Sand fraction							Clay fraction						
		Am	Gb	Fd	Cr	Qz	Py	Kao	Ill	Chl	HIV	Kao	Gb	Cr	Qz
O/A1	0–15	+	+	++	+	++	++	−	+	+	−	++	++	+	+++
A2	15–30	++	+	++	+	++	+	−							
Bw1	30–48	++	+	++	+	++	+	−	+	+	−	++	++	++	+++
Bw2	48–83	++	+	+++	+	++	+	−							
Bw3	83–110	+	+	++	+	++	++	−							
BC	110–140	++	+	+++	+	++	++	−	+	+	+	++	+++	+	++
C	>140	++	+	+++	+	++	+	−+	+	+	+	++	+++	+	+++

Sand fraction: *Am* amphibole, *Gb* gibbsite, *Fd* feldspar, *Cr* cristobalite, *Qz* quartz, *Py* pyrite, *Kao* kaolinite
Clay fraction: *Ill* illite, *Chl* chlorite, *HIV* hydroxyl-interlayered vermiculite, *Kao* kaolinite: *Gb* gibbsite, *Cr* cristolbalite, *Qz* quartz
− non-detectable; + minor; ++ moderate; +++ dominant
Abstracted from Chen et al. (1999)

Table 4.8 Bulk soil and clay mineralogy of Hapludands, Andic Dystrudepts, and Typic Dystrudepts

Horizon	Bulk soil						Clay			
	7 Å	Layer silicates	Quartz	Hematite	Cristobalite	Gibbsite	Illite	Kaolin minerals	Chlorite	Gibbsite
Hapludands										
A1	tr	++	+	−	+	+	tr	+	++	++
A2	tr	+	+	−	+	+	tr	+	++	++
Bw1	tr	+	+	−	+	+	tr	+	+	++
Bw2	tr	+	+	−	+	++	tr	tr	+	++
C	−	−	tr	−	+	+++	tr	−	tr	++
Andic Dystrudepts										
A	tr	+	+	+	tr	+	+	tr	++	++
AB	tr	++	+	+	tr	+	+	tr	++	++
BA	tr	++	+	tr	tr	+	+	tr	+	++
Bw1	tr	++	+	tr	tr	+	+	tr	+	++
Bw2	tr	++	+	tr	tr	+	+	tr	++	++
Bw3	tr	++	+	tr	tr	+	+	tr	+	++
BC	tr	++	+	tr	tr	+	tr	tr	+	++
C	tr	++	+	tr	tr	+	+	tr	+	++
Typic Dystrudepts										
A	++	++	++	+	+	+	++	+++	tr	++
Bw1	++	++	+	+	+	+	++	+++	tr	++
Bw2	+++	++	+	+	+	+	+	+++	tr	+
Bw3	++	++	+	+	+	+	tr	+++	−	++
Bw4	++	++	+	+	+	+	+	+++	tr	++
Bw5	++	++	+	+	+	+	+	+++	tr	+
C	+	++	tr	+	++	++	tr	+++	−	+

tr trace amounts, + low, ++ medium, +++ high, − not detected
Abstracted from Tsai et al. (2010)

andesite and olivine bearing hornblende-two pyroxene andesite (Chang 1974), and basalt and agglomerate (Chang 1975) were selected for examining clay mineralogy and related chemical properties. The variations of soil morphology, physical and chemical properties, and clay mineralogy distribution under different soil environments were compared in those studies (Table 4.9).

The results of those early studies showed that including Andisols, Inceptisols and Ultisols are the major soil groups formed on those volcanic parent materials, but except the Vertisols formed on basalt parent material. The differences of clay minerals of the studied soils are not considered as similar, but in the relative contents of the clay minerals. Chang (1972) suggested that topography and climatic factors have much influence on the chemical composition of the soils developed from volcanic detritus and two-pyroxene andesite. For soils formed on two pyroxene-hornblende andesite and olivine bearing hornblende-two pyroxene andesite, podzolization is a principal soil formation process (Chang 1974), and laterization is a principal soil formation

Table 4.9 Clay mineralogy and soil classification of different volcanic parent materials in the region of Tatun Mt

Volcanic parent material	Clay minerals	Soil classification[a]
Agglomerate	kaolinite, illite, gibbsite	Ultisols
Basalt	kaolinite, illite, gibbsite	Vertisols
Olivine bearing hornblende-two pyroxene andesite	kaolinite, illite, vermiculite (chlorite)	Ultisols
Two pyroxene andesite	Illite, vermiculite, kaolinite, gibbsite	Ultisols, Andisols or Inceptisols
Two pyroxene-hornblende andesite	Illite, kaolinite, beidellite	Ultisols
Volcanic detritus	kaolinite, illite	Andisols or Inceptisols

[a] Soil classification was based on Soil Survey Staff (1999)
Summarized from Chang (1972, 1974, 1975)
Copyright by Prof. Tsai

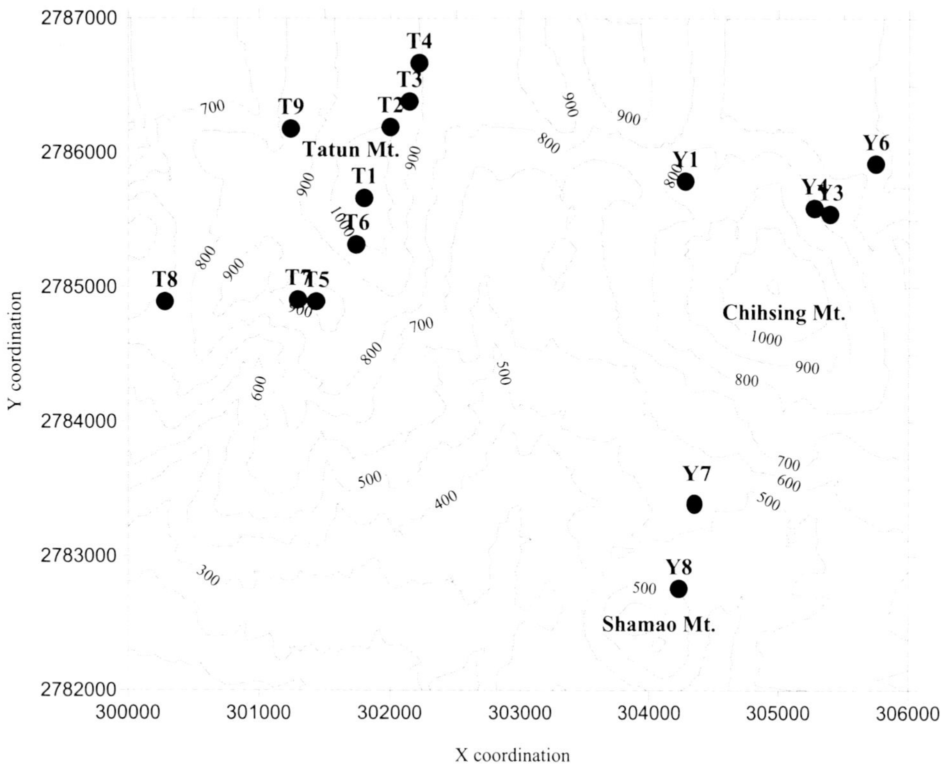

Fig. 4.6 Location of selecting soil pedons in Shamao Mt., Chihsing Mt., and Tatun Mt. Copyright by Prof. Tsai

process in soils formed from basalt and agglomerate (Chang 1975).

During 1990–1994, 15 soil pedons derived from volcanic rocks in Tatun Mt., Chihsing Mt., and Shamao Mt. were selected (Fig. 4.6) (Huang and Chen 1990; Chen and Huang 1991; Huang and Chen 1992; Huang et al. 1993, 1994). Among the pedons, 12 pedons were classified as Andisols, and 3 pedons were Inceptisols (Table 4.10) (Huang 1995). Additional, Huang (1995) indicated that significant correlations could be found among pyrophosphate extractable Al (Al_p) or Fe (Fe_p) (%), organic carbon (%), pyrophosphate extractable C (C_p) and fulvic acid (C_f), and such results suggested that the accumulation of humus in Andisols was proceeded simultaneously by the formation of Al/Fe-humus complexes. Fulvic acid was the principle humus complexed with aluminum to form Al-humus complexes. The presence of humic acid not only inhibits the formation of allophane-like minerals, but also affected the formation of crystalline Fe oxyhydroxide minerals. Short-range-ordered ferrihydrite minerals were the principle Fe-content minerals in these soils.

The differences in chemical and physical properties of these soils were attributed to the stage and process of weathering. The results of forward study in north part of Chihsing Mt. indicate that two studied soils (pedon Y1 and Y6) were classified as Andisols (Huang and Chen 1990), and amorphous materials was predominant clay minerals, and few poor crystalline aluminosilicate mineral was also found in the

X-ray diffraction (XRD) pattern. Some allophone-like and imogolite-like minerals were found in the clay fraction by TEM, and about 20 % proto-imogolite allophone in the clay fraction was estimated by the content of Si extracted by oxalic acid (Huang and Chen 1990). The climate is temperate and moist in this area. The umbric epipedon is the distinguishing feature in the surface soils of this region. Low soil pH and andic soil properties were strongly influenced by the northeastern seasonal monsoon which brings high precipitation (more than 4,000 mm/year) on soils. In the northeastern part of Chihsing Mt., microrelief has great effect on the soil characteristics and soil distribution (pedon Y3 and Y4) (Huang and Chen 1992). A large amount of organic materials has accumulated on the surface horizons, resulting in the formation of the umbric epipedon that is distinguishing feature in the field. Quartz and cristobalite are both principal components of clay minerals in pedon Y3 and Y4, and some amorphous material was estimated. As proposed by Saigusa et al. (1991), the Al_p/Al_o ratio can be used to approximately separate allophonic Andisols (ratio < 0.5) and nonallophanic Andisols (ratio $\geq$ 0.5). Among five soil pedons (Y1, Y3, Y4, Y6, and Y7), only pedon Y6 was allophonic Andisols. In addition, allophone did not exist in soil when the soil pH (H_2O) < 4.9 (Shoji et al. 1982, 1993). The soil pH of pedon Y6 was higher than 4.9 and the Al_p/Al_o ratio was less than 0.5; both meet the criteria of allophonic Andisols. The other soil pedons were non-allophanic Andisols because of lower soil pH and higher Al_p/Al_o ratio.

Table 4.10 Andic soil properties and classification of selected soil pedons in Chihsing Mt. and Tatun Mt. region

Pedon no.	Andic soil properties[a]			Soil classification[b]	Allophanic andisols[c]
	Bulk density ≤ 0.9 (Mg m^{-3})	$Al_o + 1/2Fe_o \geq 2$ %	P-retention ≥ 85 %		
Y1	O[d]	O	O	Hapludands	N
Y3	O	O	O	Hapludands	N
Y4	O	O	O	Hapludands	N
Y6	O	O	O	Hapludands	Y
Y7	O	O	O	Hapludands	N
Y8	O	O	O	Melanudands	N
T1	O	O	O	Hapludands	N
T2	O	O	O	Hapludands	Y
T3	O	O	O	Hapludands	N
T4	O	O	O	Hapludands	Y
T5	X	X	X	Dystrudepts	N
T6	O	O	O	Hapludands	N
T7	O	X	O	Dystrudepts	N
T8	O	X	O	Dystrudepts	N
T9	O	O	O	Hapludands	N

[a] Have andic soil properties throughout the subhorizons, which have an accumulative thickness of 60 % or more within 60 cm of the mineral soil surface

[b] Based on Soil Survey Staff (1992)

[c] Based on the Saigusa et al. (1991). ($Al_p/Al_o < 0.5$ allophanic Andisols; $Al_p/Al_o \geq 0.5$ nonallophanic Andisols). *N* is not allophanic Andisols, *Y* is allophanic Andisols

[d] *O* meet the criteria of andic soil properties; *X* can not meet the criteria of andic soil properties

Abstracted from Huang (1995)

Only one pedon (pedon Y8) was selected from Shamao Mt., and also an only one Andisols with melanic epipedon (Chen and Huang 1991). The parent material was volcanic detritus formed about 0.27 M years ago. The vegetation was broad-leaf mixed forests. The soil temperature regime (STR) was thermic and the soil moisture regime (SMR) was perudic. Pedon Y8, with lower pH (4.1–4.5) and higher Al_p/Al_o ratios (0.7–1.0), was classified as the nonallophanic Andisols. The characteristics of pedon Y8 were listed as: thickness of melanic epipedon is more than 45 cm; black Munsell color (2.5Y 2/0); low bulk density (0.37–0.43 Mg m^{-3}); very friable consistency; high organic carbon (10.9–15.6 %); high exchangeable aluminum (3.8–8.0 cmol(+) kg^{-1} soils); high phosphate retention (>99 %); low melanic index (<1.65); high oxalate extractable Al and Fe ($Al_o + 1/2Fe_o = 2.58$–3.55 %); aluminum was mostly complexed with organic matter; only about 5 % allophane in the melanic epipedon; and clay mineral of the parent materials was predominantly imogolite-like.

Four pedons (pedons T1–T4) were sampled from northeastern Mt. Tatun, and there were little differences among the soil pedons (Table 4.10) (Huang et al. 1993). Under strong leaching conditions, the surface horizon is extremely acidic (pH 4.3–5.0, mostly $\leq$ 4.9), and Al- and Fe-humus complexes are the principal amorphous materials. The soils of subsurface horizons abound with inorganic amorphous materials, such as allophone. There are some 2:1 type aluminosilicate minerals in the clay fraction of the surface horizons, and that, contrary to the distribution of the gibbsite, the amount of aluminosilicates decreases with increasing of the soil depth. Huang et al. (1993) suggested that pedons T1–T4 are typic non-allophanic Andisols because of strong leaching processes. After reviewing and examining the soil data, we proposed that pedons T2 and T4 are allophonic Andisols based on the criteria described above. The other five soil pedons (pedon T5–T9) were located between Tatun Mt. and Meintein Mt. Under similar weathering conditions, the pedons with different volcanic lava materials have different properties and can be classified into different Soil Order (Huang et al. 1994). This study area has temperate and humid conditions. Both umbric epipedon and cambic horizon are the most diagnostic horizon of the pedons. Due to the strong leaching conditions, all the pedons display strongly acidic reactions and do not meet the andic soil properties. The clay minerals mainly contain the resistant minerals, only a small amount of allophone minerals were found in the lower part of the pedon (<5 %). Within five soil pedons, only two soils are classified as typic Andisols, and three of them are classified as Inceptisols.

Further, through the micromorphological approach, information on the nature of andesite weathering under the prevailing perhumid subtropical climate of northern Taiwan was studied (Asio and Chen 1998). The investigation was conducted at Liutzunan (LTN) of Chutzu Mt., one of the volcanic mountains in Yangmingshan Volcanic Nation Park. The study pedon was located at the summit with elevation of 860 m a.s.l., and the parent material of the soils is andesite pyroclastic of Pleistocene origin. Vegetation is dominated by kunishi cane (*Sinobambusa kunishii*). The selected soil pedon (Typic Hapludand) is medium textured, soft, porous, acidic soil (pH < 5.0), and has very high organic matter content (>8 %) in its upper horizon. The results of Asio and Chen (1998) proposed that the weathering of andesite is rapid under the prevailing environmental conditions (perhumid and well-drained) in this study area. The ferromagnesian minerals weathered ahead of feldspar. Four approximate stages of the weathering of andesite were proposed by micromorphological approaches. The rapid weathering and leaching processes observation to explain the widespread occurrence of acidification and plant nutritional problems such as Ca and P deficiencies and Al toxicity occurred in the study area.

In Meintein Mt., three pedons representing the summit (elevation 970 m, 12 % slope), back slope (elevation 890 m, 25 % slope), and foot slope (elevation 800 m, 10 % slope) position of a volcanic landscape were selected from the southeastern part (Fig. 4.7) (Chen et al. 1999). Results pointed that the studied soils were formed in their middle stage (<0.3 M years) of development and have characteristics that are closely comparable except solum depth. They showed andic properties and strongly acidic, with very high aluminum saturation. The soils are classified as Alic Fulvudands based on Keys to Soil Taxonomy or classified as Alic Andisols based on WRB system (Chen et al. 1999). The noncrystalline forms of Fe and Al are largely associated with humus, which is related to little or no allophone formation in the A horizon. Kaolinite, halloysite, gibbsite, and quartz dominate the clay minerals composition, with minor amounts of 2:1 type silicate clays. Solum thickness seemed to be related to the position in the toposequence. This difference is affected by the soil slopes of the landscape position.

In Huangtzuei Mt., three pedons representing the summit (elevation 910 m, <5 % slope), back slope (elevation 830 m, 55 % slope), and foot slope (elevation 780 m, 28 % slope) position of a volcanic landscape were examined and sampled to explain the formation of Inceptisols derived from the pyroclastic material (Fig. 4.8) (Chen et al. 2001). The results indicated that the soils were moderately developed and characterized by an A-BA-Bw-BC horizon sequence, loam to clay loam texture, low bulk density, and high water and

Fig. 4.7 The location of three soil-sampling sites in Meintein Mt. Copyright by Prof. Tsai

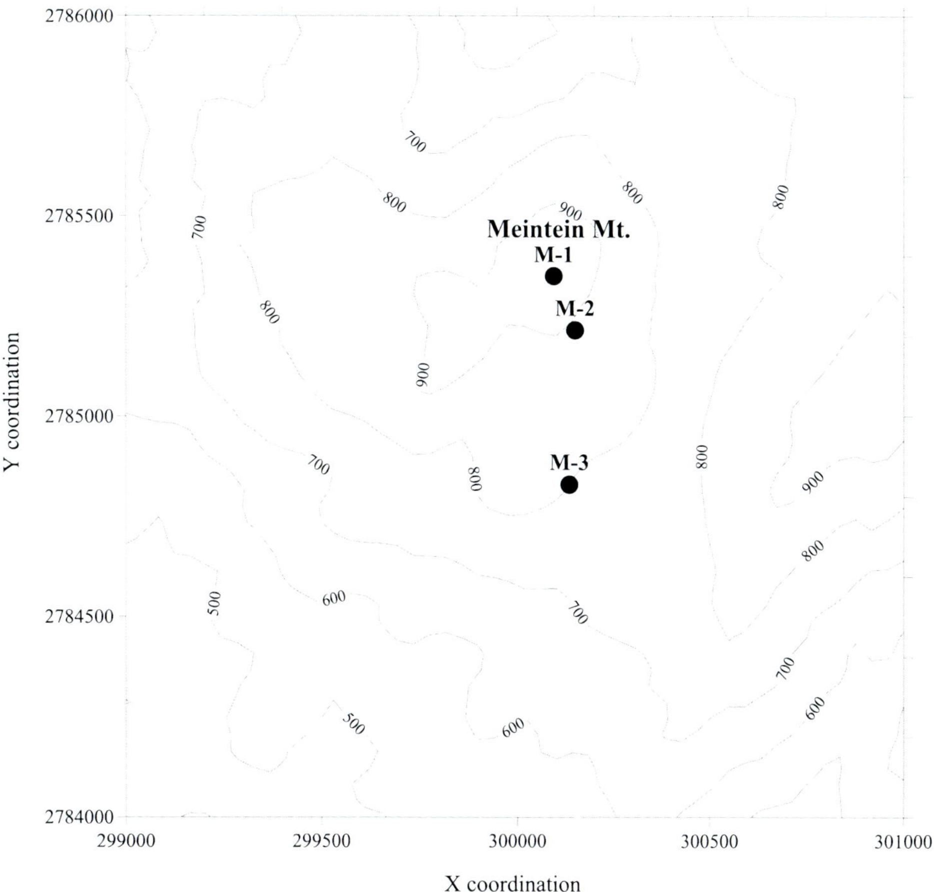

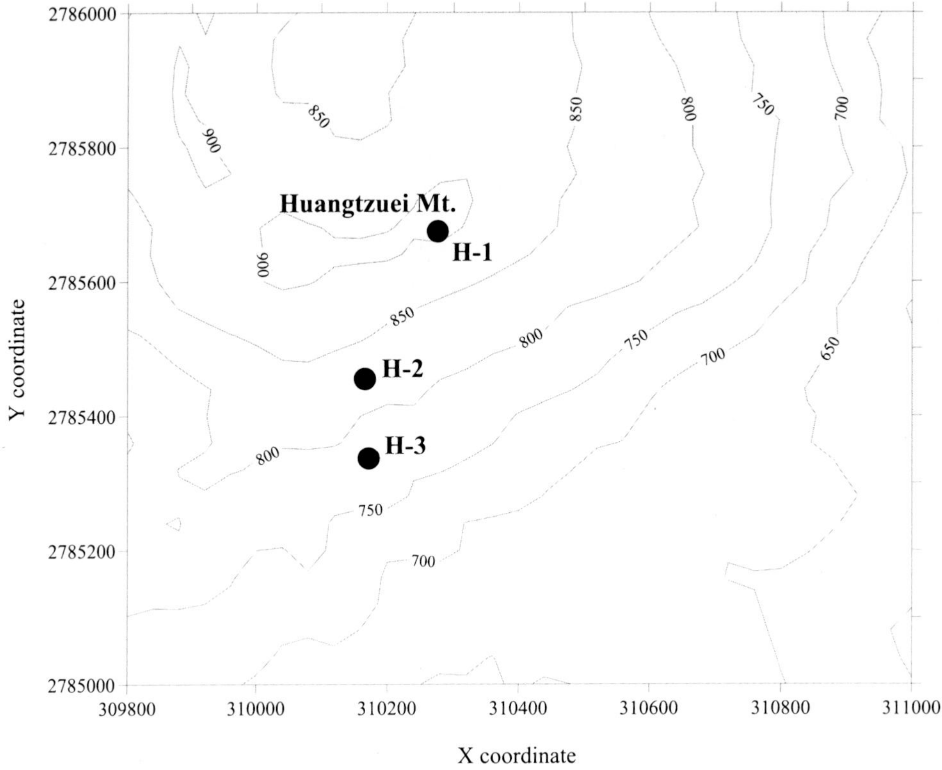

Fig. 4.8 The location of three soil-sampling sites in Huangtzuei Mt. Copyright by Prof. Tsai

P retention. All studied soils are classified as Andic Dystrudepts based on Keys to Soil Taxonomy or classified as Dystric Cambisols based on WRB system (Chen et al. 2001). Gibbsite and quartz were dominant in the sand fraction, whereas gibbsite, kaolinite, and 2:1 type silicate minerals were the major minerals in the clay fraction. Formation of the studied soils was relatively fast because of the strong weatherable nature of the pyroclastic materials, perhumid climate, and good drainage of the volcanic landscape. Therefore, it is likely that the sequence of soil formation was from Entisols to Andisols and then to Inceptisols. The most important soil-forming processes in the formation of the Inceptisols were likely structure formation, loss of bases and acidification, braunification, bioturbation, organic matter accumulation, weathering, and clay mineral formation.

Recently, the study of mineral weathering and soil formation along a soil climosequence formed on volcanic deposits in Ttatun Mt. has conducted by Tsai et al. (2010). The results show that key soil properties, such as phosphate retention, Al_o + $1/2Fe_o$, Si_o, and the Fe_o/Fe_d ratio significantly correlated with elevation. Similar altitudinal trends were reported by Zehetner et al. (2003) in the inter-Andean valley of northern Ecuador. The observed altitudinal trends largely correspond with climatic differences as determined by elevation on the volcano. Higher rainfall and lower evapotranspiration (ET) caused by lower temperature and higher humidity at higher elevation, result in greater leaching. With decreasing elevation, precipitation decreases and ET increases, leaving less water for leaching. Additionally, the high-leaching environment at elevations above 900 m asl has resulted in the formation of andic soil properties, and the soils classify as Andisols/Andosols. The increased accumulation of organic matter has led to the formation of thick, dark (umbric) epipedons at these elevations. The abundance of organic ligands in these surface horizons may have impeded the precipitation of allophane by competition with othosilicic acid for soluble Al. As leaching decreases with decreasing elevation, andic soil properties become less strongly expressed (between 400 and 500 m a.s.l.), and disappear completely (below 300 m a.s.l.). At intermediate elevations, the soils classify as Andic Inceptisols/Andic Cambisols, and at low elevations, the soils classify as Typic Inceptisols/Haplic Cambisols. Climate is considered the overriding factor responsible for the observed altitudinal differences in soil development by affecting the weathering regimes along the studied volcanic slopes. In addition to the altitudinal changes of climatic conditions, slope and aspect may also affect local pedogenesis. In conclusion, the high-elevation soils (>900 m a.s.l.) contained metastable poorly crystalline materials (e.g. allophane, ferrihydrite) and classified as Andisols, whereas the low-elevation soils (<300 m a.s.l.) contained thermodynamically more stable minerals (e.g. kaolinite, hematite) and classified as Inceptisols. The soils at intermediate elevation (400–500 m a.s.l.) were at the threshold of andic soil properties and classified as Andic Inceptisols.

4.3 Land Use and Management

The natural vegetation in the Yangmingshan National Park had undergone several destructive activities during the last two decades, and vegetation changes in the area were highly correlated to production activities (Wang et al. 2003; Hseu et al. 2008). Grazing and productions of *Baphicacanthus cusia*, teas, Ditto rice, oranges, vegetables and flowers had contributed to the vegetation changes in various degrees. Among them, the prosperous tea production during 1875–1942 had seriously destroyed the vegetation. Forestation was also a contributing factor for vegetation composition (Wang et al. 2003). During 1895–1945, reforestations were conducted at Mt. Tatun, Chihsing, Tsaikungkeng, Mientein and Shamao. As a result, totally 2,680 ha of plantations were reforested, mainly by *Pinus massoniana*, *Pinus luchuensis*, and *Pinus thunbergi*. However, after illegal cutting and natural succession following the end of the 2nd World War, forestations on the mountain were gone, and the trace of forestation can only be identified by some remaining trees within the secondary forests. Results of vegetation analysis showed that the man-made forests in the YMS-NP can be classified into four vegetation types, namely (1) *Cinnamomum camphora* and *Ficus microcarpa* Association, (2) *Acacia confusa* and *Liquidambar formosana* Association, (3) *Cryptomeria japonica* Association, and (4) *Pinus luchuensis* Association (Wang et al. 2003).

The major vegetation changes during the last three decades include changes in agricultural forms, forest recovery on abandoned farmlands along the eastern and southern boundaries of the park, and housing development on the west side of the YMS-NP (Hseu et al. 2008). After the establishment of YMS-NP, the production activities reduced significantly. Although the original vegetation had undergone massive destruction, fortunately the destruction did not last a very long time from the point of view of vegetation succession. Current vegetation conditions suggest that succession is starting over again. The damaged lands are recovering and the forests will reach the climax in the near future.

References

Asio VB, Chen ZS (1998) Study of andesite weathering in northern Taiwan using micromorphological approaches. Taiwan J For Sci 13:259–269

Chang CM (1972) Mineralogy and related chemical properties of the soils developed from volcanic detritus and two pyroxene andesite in the Tatunshan region. Mem Coll Agric Natl Taiwan Univ 13:119–136 (in Chinese, with English summary, tables and figures)

Chang CM (1974) Mineralogical and related chemical properties of the soils developed from two pyroxene-hornblende andesite and olivine bearing hornblende-two pyroxene andesite in Tatunshan region. J Chin Agric Chem Soc 12:36–53 (in Chinese, with English summary, tables and figures)

Chang CM (1975) Clay mineralogy and related chemical properties of the soils developed from basalt and agglomerate in the Tatunshan region. J Chin Agric Chem Soc 13:46–57 (in Chinese, with English summary, tables and figures)

Chang YT, Huang S (2001) Long-term ecological research scheme in Yangmingshan national park. Commissioned research project report, Yangmingshan national park headquarters, construction and planning agency. Ministry of the Interior, Taiwan (in Chinese, with English abstract)

Chen CH (1990) Igneous rocks of Taiwan. Central Geology Institute, Ministry of Economics, ROC (in Chinese, with English summary)

Chen ZS, Huang JH (1991) Characteristics and clay minerals of andisols with melanic epipedon in Taiwan. J Chin Agric Chem Soc 29:415–426 (in Chinese, with English abstract, tables and figures)

Chen WK, Tsai CY (1983) The climate in Yangmingshan national park. Department of Atmospheric Science. National Taiwan University, Taipei (in Chinese)

Chen CH, Wu YJ (1971) Volcanic geology of the Tatun geothermal area, northern Taiwan. Proc Geol Soc China 14:5–20

Chen ZS, Asio VB, Yi DF (1999) Characteristics and genesis of volcanic soils along a toposequence under a subtropical climate in Taiwan. Soil Sci 164:510–525

Chen ZS, Tsou TC, Asio VB (2001) Genesis of inceptisols on a volcanic landscape in Taiwan. Soil Sci 166:255–266

Childs CW, Matsue N, Yoshinaga N (1991) Ferrihydrite in volcanic ash soils of Japan. Soil Sci Plant Nutr 37:299–311

Dixon JB (1989) Kaolin and serpentine group minerals. In: Dixon JB, Weed SB (eds) Minerals in soil environments, 2nd edn. SSSA Book Series 1. SSSA, Madison, WI

Ho CS (1988) An introduction to the geology of Taiwan: explanatory text of the geologic map of Taiwan, 2nd edn. Central Geological Survey, The Ministry of Economic Affairs, Taiwan (in Chinese, with English abstract, tables and figures)

Hseu LD, Wang IJ, Li TM (2008) Vegetation changes in Yangmingshan national park. Commissioned research project report, Yangmingshan national park headquarters, construction and planning agency. Ministry of the Interior, Taiwan (in Chinese, with English abstract)

Huang JH (1995) Properties of Al/Fe-humus complexes in soils derived from volcanic rocks. J Chin Agric Chem Soc 33:424–435 (in Chinese, with English abstract, tables and figures)

Huang JH, Chen ZS (1990) Characteristics, genesis, and classification of two Andisols in Chihsingshan mountain area, northern Taiwan. J Chin Agric Chem Soc 28:135–147 (in Chinese, with English abstract, tables and figures)

Huang JH, Chen ZS (1992) Soil properties and classification of Andisols in northeastern Chihsing mountain, Taiwan. J Chin Agric Chem Soc 30:216–228 (in Chinese, with English abstract, tables and figures)

Huang JH, Chen ZS, Wang MK (1993) Soil properties and clay mineralogy of Andisols in the northeastern Tatun mountain, Taiwan. J Chin Agric Chem Soc 31:325–339 (in Chinese, with English abstract, tables and figures)

Huang JH, Chen ZS, Wang MK (1994) The properties and classification of pedons derived from volcanic lava materials between Tatun mountain and Mentain mountain. J Chin Agric Chem Soc 32:294–308 (in Chinese, with English abstract, tables and figures)

Kao CI (2003) The pedogenic process and indices of transition soils between Andisols and Ultisols in northern Taiwan. Master thesis, National Taiwan University, Taipei (in Chinese with English abstract)

Lair GJ, Zehetner F, Hrachowitz M, Franz N, Maringer FJ, Gerzabek MH (2009) Dating of soil layers in a young floodplain using iron oxide crystallinity. Quat Geochronol 4:260–266

Mizota C, van Reeuwijk LP (1989) Clay mineralogy and chemistry of soils formed in volcanic material in diverse climatic regions. Soil monograph 2, ISRIC, Wageningen, The Netherlands

Nanzyo M (2003) Unique properties of volcanic ash soils. Glob Env Res 6:99–112

Parfitt RL, Wilson AD (1985) Estimation of allophane and halloysite in three sequences of volcanic soils, New Zealand. In: Fernandez Caldas E, Yaalon DH (eds) volcanic soils, suppl 7. Catena

Schwertmann U (1985) The effect of pedogenic environments on iron oxide minerals. In: Stewart BA (ed) Advances in soil science, vol 1. Springer, New York, pp 171–200

Saigusa M, Matsuyama N, Honna T et al (1991) Chemistry and fertility of acid Andisols with special reference to subsoil acidity. In: Wright RJ et al (eds) Plant-soil interactions at low pH. Kluwer Academic Publishers, Netherlands

Shoji S, Fujiwara Y (1984) Active aluminum and iron in the humus horizons of Andosols from northeastern Japan: their forms, properties, and significance in clay weathering. Soil Sci 137:216–226

Shoji S, Fujiwara Y, Yamada I et al (1982) Chemistry and clay mineralogy of ando soils, brown forest soils, and podzolic soils formed from recent Towada ashes, northeastern Japan. Soil Sci 133:69–85

Shoji S, Nanzyo M, Dahlgren R (1993) Volcanic ash soils: genesis, properties and utilization. Elsevier Publisher, Development in Soil Science #21. Amsterdam, London, New York, and Tokyo

Soil Survey Staff (1992) Keys to soil taxonomy, 5th ed. SMSS technical Monograph No. 19, Virginia Polytechnic Institute and State University, Blacksburg

Soil Survey Staff (1999) Soil Taxonomy: a basic system of soil classification for making and interpreting soil surveys, 2nd edn. USDA, Natural resources conservation service, agriculture. Handbook no. 436. Washington, DC

Soil Survey Staff (2006) Keys to soil taxonomy, 10th edn. USDA-Natural Resources Conservation Service, Washington

Soil Survey Staff (2010) Keys to soil taxonomy, 11th edn. US Dept Agric Nat Res Cons Serv, Washington

Song SR, Tsao SJ, Lo HJ (2000) Characteristics of the Tatun volcanic eruptions, north Taiwan: implications for a cauldron formation and volcanic evolution. J Geol Soc China 43:361–378

Tsai CC, Chen ZS, Kao CI et al (2010) Pedogenic development of volcanic ash soils along a climosequence in northern Taiwan. Geoderma 156:48–59

Vacca A, Adamo P, Pigna M et al (2003) Genesis of tephra-derived soils from the Roccamonfina volcano, south central Italy. Soil Sci Sco Am J 67:198–207

Wada K (1989) Allophane and imogolite. In: Dixon JB, Weed SB (eds) Minerals in soil environments, 2nd edn. SSSA Book Series No. 1, SSSA, Madison

Wang IJ, Hseu LD, Lin MY et al (2003) Long-term ecological research in Yangmingshan national park: survey of vegetation change and succession. Commissioned research project report, Yangmingshan national park headquarters, construction and planning agency. Ministry of the Interior, Taiwan (in Chinese, with English abstract)

Zehetner F, Miller WP, West LT (2003) Pedogenesis of volcanic ash soils in Andean Ecuador. Soil Sci Soc Am J 67:1797–1809

5.1 General Information

Entisols, a soil with slight development in which the properties are determined largely by the parent material, is important in rural, hill, and forest lands in Taiwan. The weak development of these soils is due to one or more of three factors (Grossman 1983): youthfulness, extremes in wetness or dryness that retarded parent material alteration, or resistance of the parent material to alternation, as for example quartzite rock. Some examples are soils in young alluvium with thin depositional layers at shallow depths, soils on hard rock, very wet mineral soils of marshes and lagoons, and soils deeply mixed by man.

5.1.1 Entisols in Rural and Hill Lands

The total area of the main island of Taiwan is approximate 36,000 km^2. Land use for agricultural production is about one-fourth of Taiwan in area, the main crop is paddy rice and the others include sugarcane, corn, vegetable, and orchard, respectively. More than 50 % of Taiwan in area is forest including broadleaf and coniferous types. Rural lands (<100 m in elevation) and hill lands (100–1,000 m in elevation) are important bases in agricultural production of Taiwan. Chen (2006) indicated that in the last 50 years, the soil survey in Taiwan have made great contribution to the soil database and good interpretation on soil management, fertilizer recommendation, and environmental quality, including detailed soil survey of Taiwan rural soils (1962–1976, 1974–1979), Taiwan hill land (1980–1988), and Taiwan forest soils in Mountains (1993–2002).

The most important and contribution of detailed soil survey of Taiwan rural soils (1962–1976, 1974–1979) is to establish 620 basic soil series of Taiwan rural soils and publish 177 soil maps at a scale of 1:25,000 for soil management and fertilization recommendation. The soil series was identified by the differences of parent material, drainage class, texture change of the profile, and soil reaction in some case. The soil survey system was followed by the Soil Taxonomy system created by USDA in 1949, and soil mapping unit is soil type for rural soils. This soil survey work was regarded as the most important and great contribution for Taiwan agricultural production and development. 620 soil series in Taiwan rural soils can be classified as nine Soil Orders based on Soil Taxonomy (see Table 1.1). The major soil orders distributed in the rural soils are Inceptisols (51.0 %), Alfisols (21.8 %), Ultisols (9.6 %), and Entisols (6.8 %). The other soil orders are only occupied in <1 % of total rural soils.

The survey area of detailed soil survey of Taiwan hill land (1980–1988) included all the hill lands with the elevation lower than 1,000 m except the rural soils which had been surveyed before 1980. Total 13 volumes of soil survey reports of whole Taiwan hill lands and 215 soil maps with the scale of 1:25,000 were published. There are 432 soil series were established by the team of soil survey in this stage. The soil series was created in this stage by the differences in parent material, drainage class, texture change of the profile, and soil reaction in some case. The soil survey system was followed by the approximate Soil Taxonomy created by USDA revised in 1960, and soil mapping unit is soil phase including surface soil texture and slope. This soil survey work was regarded as one of the most important and great contribution for Taiwan agricultural production and development.

Hseu et al. (2008) summarized that although 620 soil series of the rural lands and 432 soil series of the hill lands have been established in Taiwan, some of their nomination overlapped. Therefore, a total of 909 soil pedons were divided into eight soil orders in Soil Taxonomy (Soil Survey Staff 2006), including Alfisols, Entisols, Histosols, Inceptisols, Mollisols, Oxisols, Ultisols, and Vertisols. However, Entisols is clearly a dominant soil order which is more than 50 % of the total soil series in the reports of detailed soil survey of Taiwan rural and hill lands (Table 5.1). Moreover, the agricultural lands were generally created on young alluvial plain and for lowland rice production, so that many soil series met fluvic or aquic Suborder in Entisols.

Z.-S. Chen et al., *The Soils of Taiwan*, World Soils Book Series,
DOI 10.1007/978-94-017-9726-9_5, © Springer Science+Business Media Dordrecht 2015

Table 5.1 Major soil taxa of great group number of Entisols in the rural and hill lands of Taiwan

Order	Number	Suborder	Number	Great group	Number
Entisols	526	Aquents	75	Endoaquent	65
				Fluvaquent	2
				Psammaquent	2
				Sulfaquent	6
		Fluvents	48	Udifluvents	14
				Ustifluvents	34
		Orthents	272	Udorthents	268
				Ustorhents	4
		Psamments	131	Udipsamments	114
				Ustipsamments	17

Cited from Hesu et al. (2008)

Additionally, Psamments can be found on some alluvial fans or near currently dominant river channels, which are characterized by high sand content (>70 %) throughout the profile.

5.1.2 Entisols in Forest Lands

The detail forest soil survey in Taiwan was conducted for 10 years by Taiwan Forestry Research Institute (TFRI) from 1993 to 2002. The forest soil survey area was >1,630,000 ha, including 37 Forest Working Circles of 8 Departments of Forest District Offices and 5 Experimental Forests (Fig. 5.1). Until now, six volumes of soil survey reports including four Forest Working Circles (Laoulongshi, Yuching, Pahsianshan and Taanshee) and two Experimental Forest of TFRI (Liukei and Taimali) and more than 100 soil maps on the scale of 1:25,000 were published. More reports and soil maps will be continuously published. The soil classification of this survey was based on *Keys to Soil Taxonomy* (Soil Survey Staff 1998), and soil mapping unit is soil phase including surface soil textures and slope. This soil survey work was regarded as the most important and great contribution for Taiwan forestry production and development in the next decade.

Nine soil orders were classified in Taiwan forest soils (Chen and Hseu 1997) (see Table 1.2). The major Soil Order distributed in the forest soils are Inceptisols (44.2 %), Entisols (35.3 %), Alfisols (10.8 %), and Ultisols (7.50 %). The other soil orders are only occupied in <1 % of total forest soils in Taiwan. Tsai et al. (2008) reorganized the six volumes of published forest soil survey reports, and showed that totally 49 soil series were rearranged and classified into 4 soil orders, 4 suborders, 4 great group, and 6 subgroup based on Keys to Soil Taxonomy (Soil Survey Staff 2006). Entisols was the second dominant soil type and has only one kind of Soil Suborder (Table 5.2).

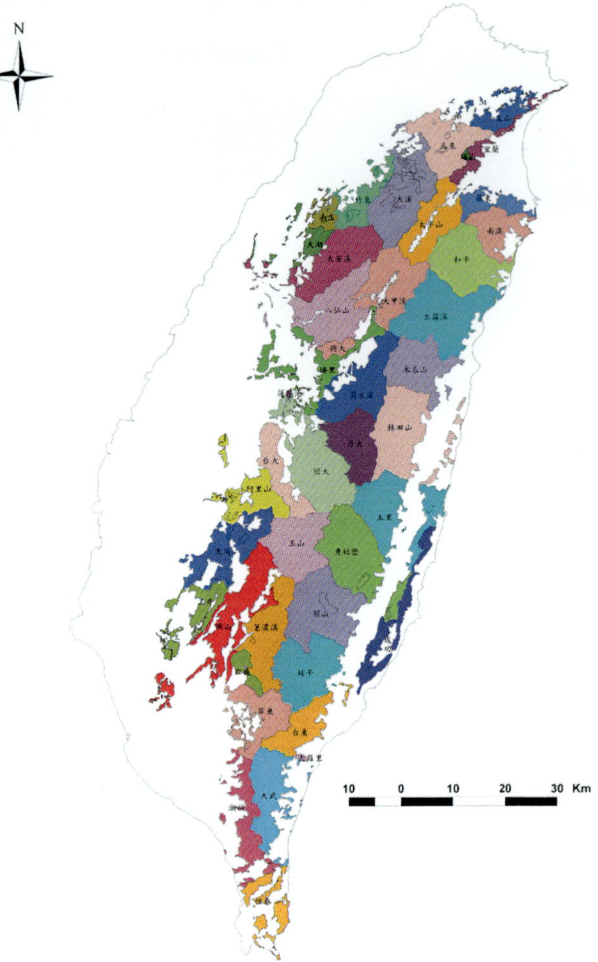

Fig. 5.1 Map of the detail forest soil survey area in Taiwan conducted by Taiwan Forestry Research Institute (TFRI) from 1993 to 2002. Copyright by Profs. Chen, Hseu, and Tsai

5.2 Profile Characteristics

5.2.1 Representative Entisols in Taiwan

For the sustainability of land use and maintenance of soil quality in Taiwan soils, Tsai et al. (1998) selected the representative soils in different counties in order to extend the agricultural technology and education on the conservation of soil resource in Taiwan. The criteria of representative soil of county are area of soil series and their contribution in agricultural production. Finally, 50 soil series are suggested to be as the Taiwan representative soils. Except Taipei, Hsingchu, and Miaoli County, 13 counties have representative Entisols (Table 5.3). In Yunglin (Central Taiwan) and Pingtung (Southern Taiwan) County, the area of representative Entisols was higher than 20,000 ha.

Among the 50 Taiwan representative soils, 20 soil series were Entisols (14 rural soil series and 6 hill land soil series) and occupied about 300,000 ha in Taiwan (Table 5.4). The

Table 5.2 Results of soil classification of 49 forest soil series

Soil order	Number	Suborder	Number	Great group	Number	Subgroup	Number
Entisols	14	Orthents	14	Udorthents	14	Lithic Udorthents	14
Inceptisols	23	Udepts	23	Dystrudepts	23	Lithic Dystrudepts	4
						Typic Dystrudepts	19
Ultisols	11	Udults	11	Hapludults	11	Lithic Hapludults	1
						Typic Hapludults	10
Spodosols	1	Orthods	1	Haplorthods	1	Typic Haplorthods	1

Based on the keys to Soil Taxonomy (Soil Survey Staff 2006)
Copyright by Prof. Tsai

Table 5.3 The representative Entisols in Taiwan County

County	Soil series	Areas (ha)	Soil great group[a]
Taoyuan	Sankengtzu	8,076	Udipsamments
Taichung	Shuipientou	5,471	Udorthents
Changhua	Yuanlin	5,575	Udipsamments
	Pingho	5,198	Udifluvents
	Shashuipu	2,000	Udipsamments
Nantou	Chechiao	1,683	Udipsamments
Yunglin	Nanhsing	10,492	Udifluvents
	NiuniaoKang	9,121	Udifluvents
	Sanliao	6,887	Udipsamments
	Hsialun	6,885	Udifluvents
Chiai	Chengchung	8,374	Ustifluvents
	Liuying	3,788	Ustifluvents
Tainan	Chengchung	17,834	Ustifluvents
Kaohsiung	Chengchung	2,100	Ustifluvents
	Funhte	2,046	Haplaquents
Pingtung	Shashuipu	24,506	Ustipsamments
	Hsipan	4,578	Ustipsamments
Penghu	Hsiaomen	785	Haplorthents
Ilan	Tayuanshe	6,236	Udorthents
	Dafu	1,010	Udifluvents
Hwalien	Juisui	12,740	Udorthents
Taitung	Juisui	12,785	Udorthents
	Chulu	2,451	Udorthents

[a] Based on keys to Soil Taxonomy (Soil Survey Staff 1996)
Cited from Tsai et al. (1998)

area of Chengchung soil series was the largest in rural soil, and Nanfu soil series was largest in hill land soil. The Suborder of representative Entisols included Orthents, Psamments, and Fluvents (Table 5.5). Udorthents, Udipsamments, Ustipsamments, Udifluvents, and Ustifluvents were the five dominant Great Groups of representative Entisols in Taiwan.

5.2.2 Characteristics of Representative Entisols

Chengchung (Cf) soil series, classified as Aquic Ustifluvents, is the largest Entisols in Taiwan agricultural soils, and widely distributes in Southern Taiwan, especially in Chiai, Tainan, and Kaohsiung County. The characteristics of Cf soil series were showed in Table 5.6. The Cf soil series has deep soil depth (130 cm), and the dominant soil color is olive brown and pale olive with some dark red and reddish brown mottles in subsoil layer (Fig. 5.2a). The soil texture was silt loam (SiL) through all horizons. The soil pH measured by soil:water = 1:1 was moderately acid in >50 cm soil depth and was neutral in <50 cm. The organic carbon (OC) was less than 10 g kg^{-1} for all soil layers, and showed gradually decrease with soil depth increased. Soil bulk density (Bd) was ranged from 1.33 to 1.57 Mg m^{-3}. The highest Bd in 20–50 cm soil layer indicated that the long-term tillage effects in Cf soil resulted in soil compaction of subsurface soil layer.

Juisui (Js) soil series, classified as Lithic Udorthents, is the second largest Entisols in Taiwan agricultural soils, and widely distributes in Eastern Taiwan, especially in Hwalien and Taitung County. The depth of Js soil series was shallow (<50 cm), and the dominant soil color is black or very dark grayish brown with yellowish red mottles in surface soil and very dark gray with strong brown mottles in subsurface soil. The soil texture was SiL or sandy loam (SL). The soil pH was mildly alkaline or moderately alkaline derived from the calcareous parent materials. The OC content ranged 10–26 g kg^{-1}, was much higher than Cf soil series. The Bd ranged from 1.12 to 1.48 Mg m^{-3}.

Shashuipu (Sp) soil series, classified as Lithic Ustipsamments, is the third largest Entisols in Taiwan agricultural soils, and widely distributes in Pingtung County, Southern Taiwan, and in Changhua County, Central Taiwan. The depth of Sp soil series is also shallow (<50 cm), and the dominant soil color is dark gray or olive gray with some dark yellowish brown mottles (Fig. 5.2b). The soil texture was

Table 5.4 The representative Entisols in Taiwan

Soil series	Areas (ha)	Soil classification (Soil Survey Staff 1996)
Agricultural soils		
Chengchung	32,526	Loamy, mixed, hyperthermic, Aquic Ustifluvents
Juisui	28,080	Sandy, mixed, hyperthermic, Lithic Udorthents
Shashuipu	26,071	Loamy, mixed, hyperthermic, Lithic Ustipsamments
Sankengtzu	11,666	Loamy, mixed, hyperthermic, Lithic Udipsamments
Nanhsing	10,492	Coarse-silty, mixed, hyperthermic, Anthraquic Ustifluvents
Hsiachung	9,601	Loamy, mixed, hyperthermic, Lithic Udipsamments
Liuying	9,463	Loamy, mixed, hyperthermic, Aquic Ustifluvents
Niuniaokang	9,121	Sandy, mixed, hyperthermic, Typic Udifluvents
Litzuyuan	7,330	Sandy, mixed, hyperthermic, Lithic Udipsamments
Chentoushan	6,732	Loamy-skeletal, mixed, hyperthermic, Lithic Udorthents
Chilu	5,768	Sandy, mixed, hyperthermic, Lithic Udorthents
Hsipan	5,135	Sandy, mixed, hyperthermic, Aquic Ustipsamments
Liuchieh	3,993	Coarse-silty, mixed, acid, hyperthermic, Lithic Udorthents
Yenchen	1,960	Loamy, mixed, hyperthermic, Typic Ustifluvents
Hill land soils		
Nanfu	42,445	Fine-loamy, mixed, hyperthermic, Lithic Udorthents
Kutingpen	30,201	Coarse-loamy, mixed, hyperthermic, Lithic Udorthents
Shihkengshan	29,356	Loamy-skeletal, mixed, hyperthermic, Lithic Udorthents
Szumahsien	14,425	Loamy-skeletal, mixed, hyperthermic, Typic Udorthents
Wutaitsun	13,963	Loamy, mixed, hyperthermic, Lithic Udorthents
Tayuanshe	6,236	Loamy-skeletal, mixed, hyperthermic, Lithic Udorthents

Cited from Tsai et al. (1998)

Table 5.5 The numbers of great groups in Taiwan representative soils based on keys to Soil Taxonomy

Soil order	Suborder	Great group	Number of soil series
Entisols	Orthents	Udorthents	10
	Psamments	Udipsamments	3
		Ustipsamments	2
	Fluvents	Udifluvents	1
		Ustifluvents	4

Soil Survey Staff (1996)
Copyright by Prof. Tsai

SiL. The soil pH was very strong acid in surface soil and strongly acid in subsurface soil. The OC content was higher (15 g kg^{-1}) in surface soil, and then decreased (8.4 g kg^{-1}). The Bd ranged from 1.31 to 1.54 Mg m^{-3}.

Sankengtzu (Sk) soil series, classified as Lithic Udipsamments, occupies large area in northern Taiwan, especially in Taoyuan County. The characteristics of Sk soil series included very shallow soil depth (<25 cm), light yellowish brown, and coarse SL texture. The soil pH was low (very strongly acid), but the OC content was high (>30 g kg^{-1}). The Bd was 1.05 Mg m^{-3} suggesting the high OC influence.

Liuying (Ly) soil series, classified as Aquic Ustifluvents, also distributes in Chiai and Tainan County. The Ly soil series has moderately deep soil depth (50–100 cm), and the dominant soil color is dark olive brown in surface soil, light olive brown and light yellowish brown in subsurface soil layers (Fig. 5.2c). The mottles with yellowish brown and very pale brown color distributed in upper 30–40 cm soil layers. The dominant soil texture was SiL. The Ly soil series has neutral to mildly alkaline soil pH. The OC content was also high in surface soil layer, and decreased with soil depth increased. In contrast, the soil Bd ranged from 1.23 to 1.65 Mg m^{-3}, and it was lower in surface layer and gradually increased with increasing soil depth. The highest Bd in 30–50 cm soil layer also suggested the long-term tillage effects in Ly soil series.

Litzuyuan (Su) soil series, classified as Lithic Udipsamments, has very shallow to shallow soil depth. The Su soil series occupied large area in Central Taiwan, including Hsingchu and Miaoli County. The dominant soil color of Su

Table 5.6 Soil characteristics of some representative Entisols in Taiwan

Soil series (code)	Depth (cm)	Munsell color (moist)	Texture[a]	pH-H$_2$O	OC (g kg^{-1})	Bd (Mg m^{-3})
Agricultural soils						
Chengchung (Cf)	0–20	2.5Y4/4	SiL	5.56	6.45	1.39
	20–50	5Y 6/4, mottle 2.5YR 3/6	SiL	5.78	5.96	1.57
	50–80	2.5Y 5/6	SiL	6.81	3.26	1.39
	80–130	5Y 6/4, mottle 2.5YR 5/3	SiL	7.27	4.35	1.33
Juisui (Js)	0–25	5Y 2.5/1, mottle 5YR 4/6	SiL	8.11	25.7	1.12
	>25	3/N, mottle 7.5YR 5/6	SiL	8.15	12.4	1.48
Juisui (Js)	0–30	2.5Y 3/2, mottle 5YR 4/6	SL	7.64	10.4	1.48
Shashuipu (Sp)	0–22	5Y 4/1, mottle 10YR 4/4	SiL	4.48	15.1	1.40
	22–40	5Y 4/2, mottle 10YR 4/4	SiL	5.41	8.42	1.54
Shashuipu (Sp)	0–20	5Y 4/2, mottle 10YR 4/6	SiL	4.48	15.1	1.31
	20–30	5Y 4/2, mottle 10YR 4/6	SiL	5.41	8.42	1.53
Sankengtzu (Sk)	0–20	2.5Y 6/4	SL	4.98	37.5	1.05
Liuying (Ly)	0–15	2.5Y 3/3, mottle 10YR 5/4	SiL	6.18	20.6	1.23
	15–40	2.5Y 5/3, mottle 10YR 7/4	SiL	6.97	15.2	1.50
	40–55	2.5Y 6/4, mottle 2.5Y 5/2	SiL	7.84	8.96	1.65
	55–70	2.5Y 5/2, mottle 2.5 Y 6/4	L	7.81	11.9	1.54
	70–100	2.5Y 6/6, mottle 5Y 6/4	SiL	7.84	9.91	1.57
Liuying (Ly)	0–15	2.5Y 3/3, mottle 10YR 5/4	SiL	7.82	11.8	1.42
	15–30	2.5Y 5/3, mottle 10YR 7/4	SiL	7.60	12.3	1.65
	30–62	2.5Y 6/4, mottle 2.5Y 5/2	SiL	7.33	9.66	1.60
	62–75	2.5Y 5/2, mottle 2.5 Y 6/4	L	7.37	7.11	1.61
	>75	2.5Y 6/6, mottle 5Y 6/4	SiL	7.52	6.70	1.57
Litzuyuan (Su)	0–30	10Y 4/2	L	7.63	30.1	1.60
Litzuyuan (Su)	0–11	5Y 3/1	SL	5.40	31.2	0.98
	11–22	5Y 3/1	SL	5.68	17.9	1.39
Chentoushan (Ct)	0–10	10Y 3-4/1	L	4.35	35.1	–
Chentoushan (Ct)	0–32	5Y 3/1	L	6.39	24.8	1.15
Chulu (Cl)	0–20	10YR 4/2	SL	7.81	12.3	1.11
Liuchieh (Lc)	0–20	5Y 4/1	SL	6.46	8.76	1.61
Liuchieh (Lc)	0–10	7.5YR 4/1	L	6.42	26.2	1.03
	10–20	10YR 4/1	L	6.32	17.8	1.08
Yenchen (Yl)	0–25	2.5Y 4/3	L	6.35	8.94	1.69
	25–45	10YR 4/4, mottle 10YR 5/8	L	7.51	4.46	1.66
	45–70	2.5YR 5/4, mottle 5Y 5/3	L	7.50	2.36	1.53
	>70	10YR 4/4	LS	7.74	1.29	–
Yenchen (Yl)	0–20	2.5Y 6/2, mottle 7.5YR 5/8 (10 %)	L	4.13	22.0	1.24
	20–40	2.5Y 4/4, mottle 7.5YR 3/4 (5 %)	SL	3.53	4.80	1.82
	40–60	2.5Y 6/6, mottle 5Y 3/1 (15 %)	LS	3.70	0.10	1.67
Shuipientou (TSp)	0–15	5Y 5/1, mottle 10YR 5/4	SL	6.57	10.1	1.56
	15–32	–	SL	6.45	5.2	1.61
Chechiao (TCo)	0–15	10YR 4/2, mottle 2.5Y 6/0	SiL	5.38	18.6	1.12
	15–30	10YR 5/4, mottle 10YR 5/1, 10YR 3/1	SiL	5.63	14.1	1.29

(continued)

Table 5.6 (continued)

Soil series (code)	Depth (cm)	Munsell color (moist)	Texture[a]	pH-H$_2$O	OC (g kg^{-1})	Bd (Mg m^{-3})
Dafu (Df)	0–20	2.5Y 4/2, mottle 7.5YR 4/6	SiL	4.80	17.1	1.26
	20–40	2.5Y 4/2, mottle 7.5YR 4/4	SiCL	4.45	18.2	1.41
	40–75	2.5YR 4/2, mottle 5YR 3/4	SiCL	5.00	16.3	1.32
	75–100	5Y 2.5/1	SiCL	6.26	16.0	1.46
Dafu (Df)	0–13	2.5Y 4/3, mottle 2.5Y 4/1, 10YR 4/4	SiL	5.33	26.6	1.19
	13–26	2.5Y 4/2, mottle 10YR 4/4	SiL	5.52	13.3	1.48
	26–56	5Y 4/1, mottle 2.5Y 5/4, 10YR 4/4	SiL	6.67	11.1	1.38
	56–70	5Y 4/1, mottle 2.5Y 5/4	SiL	7.08	11.5	1.36
	70–100	5Y 4/1, mottle 2.5Y 5/4	LS	7.07	6.18	1.41
Hill land soils						
Nanfu (Naf)	0–15	10YR 5/4	CL	6.17	7.70	1.60
	15–45	10YR 5/4	CL	6.48	0.48	1.5
Kutingpen (Kip)	0–20	10YR 5/4	L	4.56	9.62	1.52
	20–40	10YR 5/6	L	4.43	5.01	1.44

Abbreviation OC organic carbon; *Bd* bulk density
[a] *L* Loam; *CL* Clay loam; *LS* Loamy sand; *SL* Sandy loam; *SiL* Silt loam; *SiCL* Silty clay loam
Copyright by Prof. Tsai

soil could be dark grayish brown or very dark gray, and no mottles was described. The soil texture was loam (L) or SL through all horizons. The soil pH could be mildly alkaline or strongly to moderately acid. The content of OC was much higher in surface soil (>30 g kg^{-1}). Also the surface soil has lower value (0.98 Mg m^{-3}) of Bd, and show dense (1.39 Mg m^{-3}) in subsurface soil.

Chentoushan (Ct) soil series, classified as Lithic Udorthents, mainly distributed in Eastern Taiwan, especially in Ilan County. Very shallow to shallow soil depth, dark gray to very dark gray soil color, and loamy soil texture are the dominant soil characteristics of Ct soil. Because of the well soil drainage, no mottles occurred in soil profile. According to the crop type and the duration time of tillage, the soil pH could be extremely acid or slightly acid. The OC was higher than 20 g kg^{-1}. Soil Bd was about 1.2 Mg m^{-3}.

Chilu (Cl) soil series, classified as Lithic Udorthents, is similar to Js soil series and mainly distributes in Hwalien and Taitung County. Very shallow depth, dark grayish brown soil color, and coarse soil texture (SL) are the dominant soil characteristics of Cl soil (Fig. 5.2d). Mildly alkaline soil pH could derive from calcareous parent material. The OC content is 12.3 g kg^{-1} and the Bd is 1.1 Mg m^{-3}.

Liuchieh (Lc) soil series, classified as Lithic Udorthents, is similar to Ct soil series and mainly distributes in Ilan County. Very shallow soil depth, dark gray, and loamy soil texture (L or SL) are the dominant soil characteristics of Lc soil. Because of the well soil drainage, no mottles occurred in soil profile. The Lc soil is slightly acid and has about 10–30 g OC kg^{-1}. Soil Bd ranged from 1.03 to 1.61 Mg m^{-3}.

Yenchen (Yl) soil series, classified as Typic Ustifluvents, is similar with Cf and Ly soil series and mainly distributes in Chiai and Tainan, and Kaohsiung County. The area of Yl soil is mall, but the soil is important agricultural soil in Southern Taiwan. The Yl soil series has moderately deep soil depth, loamy soil texture in surface, and coarser loamy texture in deep soil layer. The surface soil has olive brown or light brownish gray soil color, and in subsurface soil the soil color changes to dark yellowish brown, light olive brown or olive yellow (Fig. 5.2e). The mottles with dark brown, strong brown, or yellowish brown color distributed in surface or subsurface soil layer. The soil pH could be extremely acid (pH 3.5–4.1) or slightly acid to mildly alkaline (pH 6.4–7.7), and those changes depend on the crop type and the duration time of tillage. Higher OC (9–22 g kg^{-1}) exist in surface soil and quickly decreased (0.1–4.8 g kg^{-1}) in subsurface soil layer. Soil Bd ranged from 1.24 to 1.82 Mg m^{-3}. The highest Bd in 20–40 cm soil layer indicated that the long-term tillage effects in Yi soil resulted in soil compaction of subsurface soil layer.

Shuipientou (TSp) soil series, classified as Udorthents, mainly distribute in Central Taiwan, especially in Taichung and Nantou County. The TSp soil is shallow with gray soil color and some yellowish brown mottles. The coarser soil texture (SL) and some mottles suggested the soil drainage could be well to somewhat poor. Slightly acid to neutral soil pH could be derived from calcareous parent materials. The OC content and Bd ranged from 5.2 to 10.1 g kg^{-1} and from 1.56 to 1.61 Mg m^{-3}, respectively. Higher Bd values of Tsp soil could be attributed to the lower OC content.

Fig. 5.2 Soil morphology of
a Chengchung (Cf) soil series,
b Shashuipu (Sp) soil series,
c Liuying (Ly) soil series,
d Chulu (Cl) soil series,
e Liuchieh (Lc) soil series,
f Yenchen (Yl) soil series.
Copyright by Profs. Chen, Hseu,
and Tsai

Chechiao (TCo) soil series, classified as Udipsamments, mainly distribute in Nantou County, Central Taiwan. Similar to TSp soil series, the depth of TCo soil is also shallow. Dark grayish brown surface soil includes some gray mottles, and yellowish brown subsurface soil includes gray and very dark gray mottles. Soil texture of TCo soil (SiL) is finer than TSp soil (SL). The OC content is higher than TSp soil, ranging from 14 to 19 g kg^{-1}, and such higher OC content could reduced the soil Bd value (1.1–1.3 Mg m^{-3}) of TCo soil.

Dafu (Df) soil series, classified as Udifluvents, mainly distribute in Ilan County, Eastern Taiwan (Table 5.5). The Df soil has moderately deep soil depth, and the dominant soil texture is SiL and silty clay loam (SiCL). The dominant soil color of Df soil is dark grayish brown or dark gray, and the color of mottles include brown, strong brown, dark reddish brown, dark yellowish brown, or light olive brown distributed (Fig. 5.2f). The finer soil texture and variable mottles suggested that the drainage of Df soil could be some what poor or poor. The OC content ranges from 11 to 27 g kg^{-1}, and decrease with soil depth increase. Soil Bd ranges from 1.2 to 1.5 Mg m^{-3}, and the highest Bd occurred in 20–40 cm soil layer.

Nanfu (Naf) soil series, classified as Lithic Udorthents, is the largest hill land soil series and is widely distributed in Taoyuan, Hsingchu, Miaoli, Taichung, Nantou, Yunglin, and Chiai County. The characteristics of Naf soil include yellowish brown soil color, shallow soil depth, finer soil texture (clay loam, CL), and slightly acid soil pH (Table 5.6). The OC content is low (<10 g kg^{-1}), especially in subsurface soil layer (<1 g kg^{-1}). Soil Bd is about 1.5–1.6 Mg m^{-3}.

Kutingpen (Kip) soil series, classified as Lithic Udorthents, is the second largest hill land soil series and widely distributes in Nantou, Yunglin, Chiai, and Tainan County. The soil characteristics of Kip soil is similar to Naf soil, including yellowish brown soil color and shallow soil depth.

Unlike the Naf soil, Kip soil has coarser soil texture (L), extremely to very strongly acid soil pH, and higher OC content (5.0–9.6 g kg^{-1}). The Bd value of Kip soil (1.4–1.5 Mg m^{-3}) is lower than Naf soil.

References

Chen ZS (2006) Pedological activities of Taiwan in last 50 years and strategies for the future. In: Proceedings of abstract of 50th Anniversary symposium of Japan society of pedology (JSP). College of Bioresources Sciences, Nihon University, Tokyo, Japan, 2 pp

Chen ZS, Hseu ZY (1997) Total organic carbon pool in soils of Taiwan. Proc Nat Sci Counc ROC Part B: Life Sci 21:120–127

Grossman RB (1983) Entisols. In: Wilding LP, Smeck NE, Hall GF (eds) Pedogenesis and soil taxonomy, vol II. Soil Orders, Elsevier, Amsterdam

Hseu ZY, Tsai CC, Guo HY et al (2008) Comparison of Taiwan soil classification system to soil taxonomy and WRB system and its modification toward a common format of Asian soil information systems. In: Proceedings of monsoon Asia agro-environmental research consortium (MARCO) workshop: a new approach to soil information systems for natural resources management in Asian Countries. Tsukuba International Congress Center, Tsukuba, Japan, pp 55–69

Soil Survey Staff (1996) Keys to soil taxonomy, 7th edn, SMSS Technical Monograph No. 19. Pocahontas Press, Inc., Blacksburg

Soil Survey Staff (1998) Keys to soil taxonomy, 8th edn. U.S. Dep. Agric, Soil Conservation Service, Washington, DC

Soil Survey Staff (2006) Keys to soil taxonomy, 10th edn. U.S. Dep. Agric, Soil Conservation Service, Washington, DC

Tsai CC, Chang YF, Hsu CW et al (2008) Proposed forest soil management divisions (FSMDs) in Taiwan. Ilan Univ J Bioresour 4 (2):99–107 (in Chinese with English abstract)

Tsai CC, Chen ZS, Hseu ZY et al (1998) Representative soils selected from arable and slope soils in Taiwan and their database establishment. Soil Environ 1(1):73–88 (in Chinese with English abstract)

6.1 General Information

Inceptisols, as contrasted with many of the other orders, include soils from a wide range of environments. However, the central concept of Inceptisols includes those soils from ustic and udic regions that have altered horizons resulting from translocation of loss of iron, aluminum, or bases. In general, Inceptisols are soils that have undergone modifications of the parent material by soil-forming processes that are sufficiently great to distinguish the soils from Entisols, but not intense enough to form the kinds of horizons that are required for classification into other soil orders (Foss et al. 1983). Most Inceptisols have cambic horizons and most are eluvial soils in that they have lost constituents by leaching and have no horizons in which significant amounts of these constituents have accumulated, but these are not necessary conditions for their recognition. Inceptisols appear to be the result of two major situations: (1) soils that are developing on geologically young sediments or landscapes; and (2) soils developing in areas where environmental conditions inhibit soil-forming processes.

6.1.1 Inceptisols in Rural and Hill Lands

As indicated in Chap. 5, 620 soil series in Taiwan rural soils can be classified as nine soil orders based on Soil Taxonomy (see Table 1.1). The major soil orders distributed in the rural soils are Inceptisols (51.0 %), Alfisols (21.8 %), Ultisols (9.6 %), and Entisols (6.8 %). In addition, a total of 909 soil series, including 620 rural soil series and 432 hill land soil series were divided into eight soil orders in Soil Taxonomy (Soil Survey Staff 2006; Hseu et al. 2008), including Alfisols, Entisols, Histosols, Inceptisols, Mollisols, Oxisols, Ultisols, and Vertisols. However, Inceptisols is the second dominant soil order in the reports of detailed soil survey of Taiwan rural and hill lands (Table 6.1).

6.1.2 Inceptisols in Forest Lands

Nine soil orders were classified in Taiwan forest soils (Chen and Hseu 1997) (see Table 1.2). The major Soil Order distributed in the forest soils are Inceptisols (44.2 %), Entisols (35.3 %), Alfisols (10.8 %), and Ultisols (7.50 %). The other soil orders are only occupied in <1 % of total forest soils in Taiwan. Tsai et al. (2008) reorganized the six volumes of published forest soil survey reports, and showed that totally 49 soil series were rearranged and classified into 4 soil orders, 4 suborders, 4 great group, and 6 subgroup based on Keys to Soil Taxonomy (Soil Survey Staff 2006). Inceptisols is the major dominant soil type and has only one kind of Soil Suborder (Table 6.2).

6.2 Profile Characteristics

6.2.1 Representative Inceptisols in Taiwan

Except Taipei, Taoyuan, Yunglin, and Penghu County, 12 counties have representative Inceptisols (Table 6.3) (Tsai et al. 1998). In Changhua County (central Taiwan), the area of representative Inceptisols was higher than 20,000 ha. Among the 50 Taiwan representative soils, 14 soil series were Inceptisols (13 rural soil series and 1 hill land soil series) and occupied about 140,000 ha in Taiwan (Table 6.4). The area of Tsochia soil series is the largest in rural soil, and Likyu soil series is the largest in hill land soil.

Z.-S. Chen et al., *The Soils of Taiwan*, World Soils Book Series,
DOI 10.1007/978-94-017-9726-9_6, © Springer Science+Business Media Dordrecht 2015

Table 6.1 Major soil taxa of great group number of Entisols in the rural and hill lands of Taiwan

Order	Number	Suborder	Number	Great group	Number
Inceptisols	251	Aquepts	39	Endoaquepts	38
				Humaquepts	1
		Udepts	170	Dystrudepts	170
		Ustepts	42	Dystrustepts	42

Cited from Hesu et al. (2008)

Table 6.2 Results of soil classification of 49 forest soil series

Soil order	No	Suborder	No	Great group	No	Subgroup	No
Entisols	14	Orthents	14	Udorthents	14	Lithic udorthents	14
Inceptisols	23	Udepts	23	Dystrudepts	23	Lithic dystrudepts	4
						Typic dystrudepts	19
Ultisols	11	Udults	11	Hapludults	11	Lithic hapludults	1
						Typic hapludults	10
Spodosols	1	Orthods	1	Haplorthods	1	Typic haplorthods	1

Based on the keys to soil taxonomy (Soil Survey Staff 2006)
Copyright by Prof. Tsai

Table 6.3 The representative soils of every county of Taiwan

County	Soil series	Areas (ha)	Soil great group (Soil Survey Staff 1996)
Hsingchu	Fangtzupo	1,080	Eutrochrepts
Miaoli	Fangtzupo	3,389	Eutrochrepts
	Fuchi	1,700	Endoaquepts
Taichung	Lilintsun	3,350	Dystrochrepts
	Tatu	2,842	Endoaquepts
Changhua	Erhlin	15,211	Epiaquepts
	Lukang	6,031	Eutrochrepts
Nantou	Lilintsun	1,573	Epiaquepts
Chiai	Tsochia	7,331	Endoaquepts
	Jente	7,183	Ustochrepts
Tainan	Annei	10,385	Ustochrepts
	Linfengying	3,500	Ustochrepts
Kaohsiung	Tsochia	3,423	Endoaquepts
	Shihtou	2,405	Endoaquepts
Pingtung	Wukueiliao	7,369	Endoaquepts
	Chiutungchiao	2,820	Epiaquepts
Ilan	Chiwulan	2,024	Epiaquepts
Hwalien	Fenglo	1,090	Epiaquepts
Taitung	Fenglo	1,154	Epiaquepts

Cited from Tsai et al. (1998)

The Suborder of representative Inceptisols included Aquepts and Ochrepts (Table 6.5). Endoaquepts, Epiaquepts, Dystrochrepts, Ustochrepts, and Eutrochrepts are the five dominant Great Groups of representative Inceptisols in Taiwan.

6.2.2 Characteristics of Representative Entisols

Tsochia (Ts) soil series, classified as Typic Ustochrepts, is the largest Inceptisols in Taiwan agricultural soils, and is widely distributed in southern Taiwan, especially in Chiai, Tainan, and Kaohsiung County. The characteristics of Ts soil series were showed in Table 6.6. The Ts soil series has moderately deep to deep soil depth (95–120 cm), and the dominant soil color is light olive brown, but some variation shows olive gray (Fig. 6.1a). The soil mottle can be found from surface to subsurface soil layer, with olive brown, olive yellow, red, or strong brown color. The soil texture is silt loam (SiL) through all horizons. The soil pH measured by soil:water = 1:1 shows neutral or mildly alkaline, but some exception indicates slightly acid. The organic carbon (OC) was less than 20 g kg^{-1} for all soil layers, and showed gradually decrease with soil depth increased. Soil bulk density (Bd) was ranged from 1.21 to 1.75 Mg m^{-3}. The highest Bd in 20–60 cm soil layer indicated that the long-term tillage effects in Ts soil resulted in soil compaction of subsurface soil layer.

Erhlin (Eh) soil series, classified as Typic Epiaquepts, is the second largest Inceptisols in Taiwan agricultural soils, and is widely distributed in central Taiwan, especially in Changhua and Yunglin County. The depth of Eh soil series was moderately deep or deep, and the dominant soil color is very dark gray or dark olivd gray in surface layer, and dark gray, olive gray, olive brown, or dark olive brown in subsurface soil (Fig. 6.1b). The mottles only appeared in lower soil layer (>60 cm), with brown or dark yellowish brown color. The dominant soil texture was SiL. The soil pH ranged from 7.42 to 8.14, belonging to mildly alkaline to

Table 6.4 The representative soils in Taiwan

Soil series	Areas (ha)	Soil classification (Soil Survey Staff 1996)
Agricultural soils		
Tsochia	30,527	Loamy, mixed, hyperthermic, typic Ustochrepts
Erhlin	21,158	Fine-silty, mixed, calcareous, hyperthermic, typic epiaquepts
Jente	14,777	Loamy, mixed, hyperthermic, typic ustochrepts
Annei	12,754	Coarse-silty, mixed, hyperthermic, aquic ustochrepts
Wukueiliao	7,369	Fine-silty, mixed, hyperthermic, typic Endoaquepts
Lukang	6,031	Fine-silty, mixed, hyperthermic, Typic epiaquepts
Yuanlin	5,575	Fine-silty, mixed, hyperthermic, Typic Endoaquepts
Pingho	5,198	Fine-silty, mixed, hyperthermic, aquic Eutrochrepts
Fangtzupo	5,075	Loamy, mixed, hyperthermic, aquic eutrochrepts
Linfengying	5,042	Fine, mixed, hyperthermic, aquic ustochrepts
Shanhua	3,874	Fine, mixed, hyperthermic, aquic ustochrepts
Fenglo	3,338	Loamy, mixed, hyperthermic, typic epiaquepts
Chiwulan	2,024	Fine-silty, mixed, hyperthermic, typic epiaquepts
Hill land soils		
Likyu	13,544	Coarse-silty, mixed, hyperthermic, lithic dystrochrepts

Cited from Tsai et al. (1998)

Table 6.5 The numbers of great groups in Taiwan representative soils based on keys to soil taxonomy

Soil order	Suborder	Great group	Number of soil series
Inceptisols	Aquepts	Endoaquepts	3
		Epiaquepts	4
	Ochrepts	Dystrochrepts	1
		Ustochrepts	4
		Eutrochrepts	2

Soil Survey Staff (1996).
Copyright by Prof. Tsai

moderately alkaline. The pH value suggests that the Eh soil could be derived from the calcareous parent materials. The OC content ranged 11.0–32.5 g kg^{-1}, was much higher than Ts soil series. The Bd ranged from 1.26 to 1.51 Mg m^{-3}.

Jente (Je) soil series, classified as Typic Ustochrepts, widely distributes in southern Taiwan, especially in Chiai County. The characteristics of Je soil similar with the Ts soil, including moderately deep to deep soil depth, light olive brown soil color (Fig. 6.1c), neutral to mildly alkaline soil pH, and soil Bd. But the soil texture of Je soil was coarser than Ts soil, including loam (L), fine sandy loam (fSL) and fine sand (fS). The OC content was lower than Ts soil series, ranging from 2.29 to 9.83 g kg^{-1}.

Annei (An) soil series, classified as Aquic Ustochrepts, widely distributes in Tainan County. The An soil series has deep soil depth. Except 55–69 cm soil layer, the characteristics of the An soil include light olive brown and olive soil color (Fig. 6.1d), SiL soil texture, moderately alkaline soil pH, and 1.75–6.59 g kg^{-1} of OC content. In 55–69 cm soil layer, much charcoal was found and resulted in very dark gray color, silty clay loam (SiCL), neutral soil reaction (pH 7.16), and much higher OC content (18.8 g kg^{-1}). The soil Bd of An soil ranged from 1.44 to 1.55 Mg m^{-3}.

Wukueiliao (Wl) soil series, classified as Typic Endo-aquepts, widely distributes in Pingtung County, southern Taiwan. The Wl soil series has moderately deep soil depth and variable soil colors, including light brownish gray and yellowish brown or gray and dark gray soil color in upper 50 cm soil layer (Fig. 6.2a). In lower soil layer (<50 cm), the soil color also shows variable. Silty loam (SiL) is the dominant soil texture class, but some coarser texture also appears in lower soil layer. The Wl soil has mildly alkaline soil pH. The OC content ranged from 4.13 to 20.8 g kg^{-1}. The Bd ranged from 1.14 to 1.65 Mg m^{-3}.

Lukang (Lu) soil series, classified as Typic Epiaquepts, widely distributes in Changhua County, central Taiwan. The depth of Lu soil series is moderately deep (about 100 cm), and the dominant soil color is dark gray and very dark grayish

Table 6.6 Soil characteristics of some representative inceptisols in agricultural soils of Taiwan

Soil series (code)	Depth (cm)	Munsell color (moist)	Texture[a]	pH-H_2O	OC (g kg^{-1})	Bd (Mg m^{-3})
Tsochia (Ts)	0–20	2.5Y 5/4	SiL	7.17	18.0	1.52
	20–60	2.5Y 5/4, mottle 2.5Y 4/4	SiL	7.37	10.7	1.75
	60–85	2.5Y 5/4, mottle 2.5Y 6/6	SiL	7.80	4.65	1.71
	85–100	2.5Y 6/6, mottle 5Y 5/2	SiL	7.70	3.79	1.64
Tsochia (Ts)	0–20	2.5Y 5/4	SiL	7.39	18.9	1.45
	20–40	2.5Y 5/4, mottle 2.5Y 4/4	SiL	7.78	10.5	1.62
	40–70	2.5Y 5/4, mottle 2.5Y 6/6	SiL	7.43	9.08	1.61
	70–100	2.5Y 6/6, mottle 5Y 5/2	SiL	7.29	7.84	1.56
Tsochia (Ts)	0–20	5Y 4/2, mottle 2.5YR 4/8 (5 %)	SiL	7.13	15.8	1.47
	20–30	5Y 3/2, mottle 7.5YR 5/8 (5 %)	SiL	7.56	12.9	1.52
	30–60	5Y 4/4, mottle 2.5YR 4/8 (40 %)	SiL	7.55	7.26	1.66
	60–90	5Y 4/4, mottle 7.5YR 4/4 (50 %)	SiL	7.44	5.49	1.61
	90–120	10YR 3/6, mottle 2.5YR 6/3 (10 %)	SiL	7.34	6.18	1.52
Tsochia (Ts)	0–15	2.5Y 6/4, mottle 2.5Y 6/8	SiL	5.39	9.15	1.26
	15–40	2.5Y 4/4, mottle 2.5Y 5/2	SiL	6.41	6.80	1.27
	40–65	2.5Y 5/4, mottle 2.5Y 5/1	SiL	6.38	6.32	1.21
	65–95	2.5Y 5/4, mottle 2.5Y 5/2, 7.5YR 5/6	SiL	5.89	4.20	1.45
	>95	2.5Y 5/1, mottle 10YR 5/8	SiCL	4.97	8.25	1.46
Erhlin (Eh)	0–20	5Y 3/1	SiL	7.48	32.5	1.26
	20–42	5Y 4/2	SiL	7.73	25.4	1.49
	42–61	2.5Y 4/3	SiL	8.14	12.0	1.50
	61–90	2.5Y 3/3	SiL	7.85	22.0	1.31
	90–150	5Y 2.5/1, mottle 10YR 3/4	SiL	8.05	11.0	1.28
Erhlin (Eh)	0–20	5Y 3/2	SiL	7.42	28.8	1.41
	20–30	5Y 4/1	SiL	7.71	25.1	1.47
	30–60	5Y 4/2	SiL	7.93	11.2	1.51
	>60	5Y 5/2, mottle 10YR 4/3	Si	7.77	11.3	1.45
Jente (Je)	0–20	2.5Y 4/6, mottle 5Y 6/4	L	6.35	8.94	1.69
	20–35	5Y 4/6, mottle 2.5Y 7/8	SiL	7.51	4.46	1.66
	35–70	2.5Y 5/4	fSL	7.50	2.36	1.53
Jente (Je)	0–20	5Y 4/2, mottle 10GY 4/1 (20 %)	L	6.74	9.83	1.50
	20–35	2.5Y 5/3, mottle 10YR 5/6 (5 %)	SL	7.94	9.56	1.75
	35–60	2.5Y 4/3, mottle 5Y 4/2 (10 %)	fS	7.68	2.35	1.65
	60–80	2.5Y 4/3, mottle 5Y 4/2 (30 %)	fS	7.55	2.42	1.58
	80–120	2.5Y 4/3, mottle 5Y 4/2 (30 %)	fS	7.42	2.29	1.56
Annei (An)	0–20	2.5Y 5/4	SiL	8.17	1.75	1.47
	20–55	2.5Y 5/4	SiL	8.09	3.77	1.53
	55–69	5Y 3/1 (charcoal)	SiCL	7.16	18.8	1.45
	69–90	5Y 4/3	SiL	7.95	6.56	1.55
	90–130	5Y 5/4	SiL	8.02	4.39	1.44
Wukueiliao (Wl)	0–25	10YR 6/2, mottle 7.5YR 4/4	SiL	7.10	20.5	1.36
	25–50	10YR 5/4	SiL	7.75	9.33	1.52
	50–60	2.5Y 4/4	LS	7.69	5.26	1.65
	60–80	2.5Y 4/4	S	7.69	5.13	1.48
	80–100	2.5Y 3/2	S	7.74	4.13	1.45

(continued)

Table 6.6 (continued)

Soil series (code)	Depth (cm)	Munsell color (moist)	Texture[a]	pH-H_2O	OC (g kg^{-1})	Bd (Mg m^{-3})
Wukueiliao (Wl)	0–30	5Y 4/1, mottle 10YR 3/3	SiL	4.85	20.8	1.25
	30–50	5Y 5/1	SiL	7.07	14.8	1.40
	50–80	10YR 5/3	SiL	7.34	9.93	1.14
	>80	2.5YR 5/2	SiL	7.40	8.10	1.34
Lukang (Lu)	0–20	5YR 4/1, mottle 2.5Y 3/4	SiL	7.79	15.1	1.43
	20–45	10YR 3/2, mottle 2.5Y 4/2	SiC	8.07	5.77	1.69
	45–80	5Y 4/4	SiC	8.05	10.1	1.66
	80–100	5Y 5/3, mottle 5Y 6/4	SiL	7.94	2.04	1.70
	>100	5Y 5/3, mottle 5Y 6/4	SiL	7.91	3.19	1.62
Fangtzupo (Fp)	0–20	2.5Y 5/1	SiCL	4.68	38.3	1.26
	20–50	2.5Y 4/1	SiCL	7.34	13.5	1.63
	50–70	2.5Y 6/1, mottle 7.5YR 5/8	LS	7.59	8.17	1.63
	70–85	2.5Y 6/2	LS	7.36	6.56	1.38
	85–120	10YR 4/1	LS	7.26	6.83	1.38
Fangtzupo (Fp)	0–20	5Y 5/1, mottle 5YR 5/8	SiL	4.90	33.1	1.20
	20–43	5Y 6/1, mottle 5YR 5/8	SiL	5.60	13.8	1.58
	43–60	5Y 6/1, mottle 5YR 5/6	SiL	6.10	4.64	1.54
	60–71	10YR 6/8, mottle 2.5Y 7/1	fSL	6.13	5.51	1.47
	71–103	10YR 6/8, mottle 2.5Y 7/1	fSL	5.82	6.80	1.31
	>103	10YR 4/1, mottle 5YR 4/6	CL	6.24	12.5	1.58
Fangtzupo (Fp)	0–23	7.5YR 3/1	SiL	7.43	1.01	1.50
	23–35	7.5YR 6/8, mottle 10YR 5/3, 2.5Y 3/1	CL	7.48	0.53	1.54
	35–50	10YR 5/3, mottle 2.5Y 6/4, 2.5Y 3/1	SiL	7.15	0.41	1.46
	50–85	7.5YR 4/6	SL	6.93	0.33	1.47
	>85	7.5YR 4/6	LS	6.92	0.32	1.56
Linfengying (Lh)	0–15	5Y 7/2, mottle 10YR 7/8	SiL	7.52	22.0	1.19
	15–30	2.5Y 4/4, mottle 10YR 6/6	SiCL	7.58	14.1	1.70
	30–45	2.5Y 4/3, mottle 2.5Y 6/3	SiCL	7.63	6.57	1.75
	45–80	5Y 5/2, mottle 2.5Y 5/4	SiCL	7.53	6.37	1.60
	80–100	5Y 5/4, mottle 2.5Y 6/6	SiCL	7.48	6.51	1.55
Linfengying (Lh)	0–15	5Y 4/2, mottle 10YR 7/8	SiL	8.29	18.5	1.35
	15–28	2.5Y 4/2, mottle 10YR 6/6	SiCL	8.51	8.47	1.82
	28–47	2.5Y 4/3, mottle 2.5Y 6/3	SiCL	8.40	6.98	1.68
Fenglo (Fl)	0–20	10Y 4/1, mottle 5YR 3/4 (10 %)	SiL	7.65	13.2	1.37
	20–40	5Y 3/1, mottle 5YR 3/4 (10 %)	SiL	7.80	10.4	1.61
	40–60	5Y 5/3, mottle 5Y 6/3 (30 %)	SiL	7.93	5.94	1.47
	60–90	5Y 6/3, mottle 2.5Y 5/6 (10 %)	SiL	7.88	4.51	1.39
	90–120	5Y 4/2, mottle 2.5Y 5/6 (30 %)	SiCL	8.03	6.18	1.30
Chiwulan (Ca)	0–16	7.5GY 4/1, mottle 5Y 6/6	SiCL	4.96	27.1	1.11
	16–25	N 4/0, mottle 5Y 6/6	SiCL	4.88	22.6	1.41
	25–60	5BG 4/1, mottle 2.5Y 6/8	SiCL	4.92	20.3	1.53
	60–70	10YR 6/1	SiCL	5.88	9.79	1.38
	>70	10YR 7/1	SiCL	5.74	8.57	1.28

(continued)

Table 6.6 (continued)

Soil series (code)	Depth (cm)	Munsell color (moist)	Texture[a]	pH-H$_2$O	OC (g kg^{-1})	Bd (Mg m^{-3})
Chiwulan (Ca)	0–17	5Y 4/1, mottle 7.5YR 4/4	SiCL	4.87	38.0	1.06
	17–42	5Y 3/1, mottle 7.5YR 4/6	SiCL	6.24	40.1	1.32
	42–62	5Y 4/1, mottle 7.5YR 5/6	SiCL	6.52	18.2	1.30
	62–110	5Y 4/1, mottle 7.5YR 5/6	SiCL	5.52	11.1	1.32
Fuchi (Fc)	0–19	10YR 5/1	L	5.92	28.5	1.20
	19–30	10YR 4/1	SL	5.74	19.7	1.70
	30–47	10YR 5/3	LS	6.87	9.67	1.65
	47–70	10YR 4/2	LS	7.28	5.98	1.51
	>70	10YR 4/1	S	7.04	4.77	1.41
Lilintsun (TLb)	0–30	5Y 6/1, mottle 10YR 5/4	SiL	6.17	13.7	1.16
	30–60	2.5Y 6/2, mottle 10YR 6/6, 10YR 3/2	SiL	7.45	5.40	1.68
	60–85	10YR 5/4, mottle 2.5Y 7/2	SiCL	7.38	4.60	–
Lilintsun (TLb)	0–20	5Y 6/1, mottle 10YR 5/4	SiL	5.37	15.6	1.24
	20–45	2.5Y 6/2, mottle 10YR 6/6, 10YR 3/2	SiL	6.16	5.27	1.74
	45–75	10YR 5/4, mottle 2.5Y 7/2	SiCL	6.73	5.85	1.62
	75–100	10YR 5/6, mottle 10YR 7/3	SiCL	6.75	4.09	1.69
Tatu (TTq)	0–25	5Y 6/1, mottle 10YR 5/4	L	5.19	3.10	1.52
	25–40	5Y 6/1, mottle 10YR 5/4	L	6.40	2.70	1.57
	40–50	2.5Y 5/4, mottle 2.5Y 5/2	fSL	6.67	2.40	–
	50–70	10YR 5/4, mottle 2.5Y 5/2	L	6.71	1.50	1.45
Shihtou (St)	0–25	5Y 3/1	SiCL	7.80	9.73	1.72
	25–55	5Y 4/1	SiCL	7.74	9.53	1.60
	55–85	5Y 5/1, mottle 10YR 6/8	SiCL	7.76	5.91	1.36
	85–105	5Y 6/2, mottle10YR 6/8	SiC	7.91	2.77	1.28
	>105	5Y 5/2	SiC	7.37	9.73	1.61

Abbreviation OC organic carbon; *Bd* bulk density

[a] *L* Loam; *CL* Clay loam; *LS* Loamy sand; *SL* Sandy loam; *fSL* Fine sandy loam; *SiC* Silty clay; *SiL* Silt loam; *SiCL* Silty clay loam

Copyright by Prof. Tsai

brown with dark grayish brown mottles in surface soil (0–45 cm), olive with pale olive mottles in subsurface soil (>45 cm) (Fig. 6.2b). The soil texture in 20–80 cm soil layer is silty clay (SiC), indicating some clay illuviation occurred. In surface (0–20 cm) and lower part (>80 cm), the soil texture is SiL. The soil pH of Lu soil is similar to An soil, showing mildly alkaline to moderately alkaline soil reaction. The OC content ranged 2.04–15.1 g kg^{-1}. The Bd of Lu soil is similar to Je soil, ranging from 1.43 to 1.70 Mg m^{-3}.

Fangtzupo (Fp) soil series, classified as Aquic Eutrochrepts, widely distributes in Hsingchu and Miaolo County, central Taiwan. The solum of Fp soil has moderately deep to deep soil depth. The surface soil has gray or dark gray soil color, and show variable soil colors in subsurface soil, including light brownish gray, brownish yellow, brown or strong brown (Fig. 6.2c). The color of mottles include strong brown, yellowish red or light gray. The soil texture class also shows variable, including SiCL, loamy sand (LS), SiL, fSL, SiL, CL, and SL. The variable soil texture could be attributed to the long-term tillage or different alluvium layer. The soil pH in surface soil indicates the very strongly acid soil reaction, but it shows slightly acid, neutral or mildly alkaline in subsurface soil layers. Inaddition, the surface soil has very high OC content (33–38 g kg^{-1}), and this can be used to explain the very strongly acid soil reaction. Except the surface soil, the OC content of Fp soil ranges from 0.32 to 13.8 g kg^{-1}. The Bd ranged from 1.20 to 1.63 Mg m^{-3}.

Linfengying (Lh) soil series, classified as Aquic Ustochrepts, widely distribute in Tainan County, southern Taiwan. The Lh soil has shallow to moderately deep soil depth, and the dominant soil color is light gray and olive gray with yellow mottles in surface soil (0–15 cm), and olive brown, dark grayish brown, and olive with light yellowish brown and light olive brown mottles in subsurface soil (>15 cm) (Fig. 6.2d).

Fig. 6.1 Soil morphology of **a** Tsochia (Ts) soil series, **b** Erhlin (Eh) soil series, **c** Jente (Jt) soil series, and **d** Annei (An) soil series. Copyright by Profs. Chen, Hseu, and Tsai

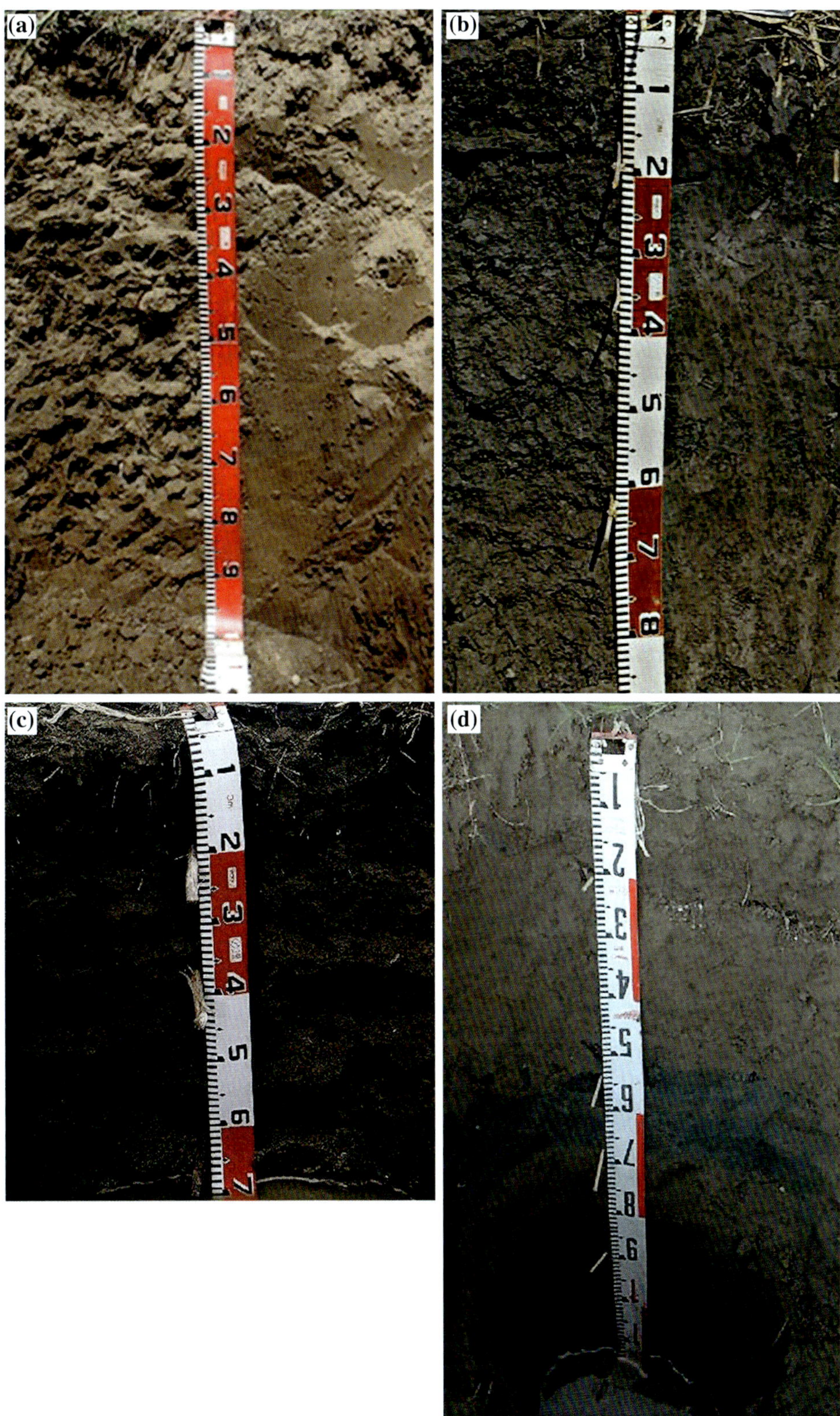

The soil texture in 0–15 cm soil layer is SiL, but it is SiCL in soil layer >15 cm, indicating some clay was leached from upper soil. The soil pH of Lh soil, showing mildly alkaline to moderately alkaline soil reaction, is similar to An and Lu soils. The OC content ranges from 6.37 to 22.0 g kg^{-1}. The Bd of Lh soil ranges from 1.19 to 1.82 Mg m^{-3}.

Fenglo (Fl) soil series, classified as Typic Epiaquepts, widely distribute in Hwalien and Taitung County, eastern

Fig. 6.2 Soil morphology of
a Wukueiliao (Wl) soil series,
b Lukang (Lu) soil series,
c Fangtzupo (Fp) soil series, and
d Linfengying (Lh) soil series.
Copyright by Profs. Chen,
Hseu, and Tsai

Taiwan. The soil characteristics of Fl soil include deep soil depth, SiL soil texture and mildly to moderately alkaline soil raction. The soil color is dark to very dark gray with dark reddish brown mottles in 0–40 cm soil layer, and olive, pale olive, and olive gary with light olive brown mottles in >40 cm soil layer (Fig. 6.3a). The OC content ranges from 4.51 to 13.2 g kg^{-1}. The Bd of Fl soil ranges from 1.30 to 1.61 Mg m^{-3}.

Fig. 6.3 Soil morphology of
a Fenglo (Fl) soil series,
b Chiwulan (Ca) soil series,
c Fuchi (Fc) soil series, and
d Shihtou (St) soil series.
Copyright by Profs. Chen, Hseu, and Tsai

Chiwulan (Ca) soil series, classified as Typic Epiaquepts, widely distribute in Ilan County, eastern Taiwan. The Ca soil has moderately deep to deep soil depth, finer soil texture (SiCL), and very strongly to slightly acid soil pH values. The dominant soil color is dark to very dark gray or greenish gray in whole soil layers, and olive yellow, brown, strong brown mottles (Fig. 6.3b). The OC contents are higher than other soil series, ranging form 8.57 to 40.1 g kg^{-1}. The Bd of Ca soil ranges from 1.06 to 1.53 Mg m^{-3}.

Fuchi (Fc) soil series, classified as Typic Endoaquepts, is the representative Inceptisols in Miaoli County, central Taiwan. The characteristics of Fc soil similar to Fp soil, including moderately deep soil depth, gray or dark gray surface soil color (Fig. 6.3c), moderately acid in surface layer and neutral in subsurface layer. The soil texture profile in Fc soil is L-SL-LS-S, also suggesting the effects of different alluviation events. The OC content ranges from 4.77 to 28.5 g kg^{-1}. The Bd ranges from 1.20 to 1.70 Mg m^{-3}.

Lilintsun (TLb) soil series, classified as Typic Dystrochrepts, is the representative Inceptisols in Nantou County, central Taiwan. The soil characteristics of TLb soil include moderately soil depth, SiL soil texture in surface soil and SiCL soil texture in subsurface soil, and slightly acid mildly alkaline soil raction. The soil color is gray and light brownish gray with yellowish brown and very dark grayish brown mottles in 0–60 cm soil layer, and yellowish brown light gray or very pale brown mottles in >60 cm soil laye. The OC content ranges from 4.09 to 15.6 g kg^{-1}. The Bd ranges from 1.16 to 1.74 Mg m^{-3}.

Tatu (TTq) soil series, classified as Typic Endoaquepts, is the representative Inceptisols in Taichung County, central Taiwan. The characteristics of TTq soil include moderately deep soil depth, gray soil color with yellowish brown mottles in surface soil layer (0–40 cm) and light olive brown and yellowish brown soil color with grayish brown mottles in subsurface soil layer (>40 cm), loamy (L) soil texture, and strongly to slightly acis soil pH. The OC contents are relative lower than other soil series, ranging form 1.50 to 3.10 g kg^{-1}. The Bd of TTq soil ranges from 1.45 to 1.57 Mg m^{-3}.

Shihtou (St) soil series, classified as Typic Endoaquepts, is the representative Inceptisols in Kaohsiung County, southern Taiwan. The characteristics of St soil include moderately deep soil depth, dark to very dark gray soil color in surface soil layer (0–55 cm) and gray, light olive gray soil color with brownish yellow mottles in subsurface soil layer (>55 cm) (Fig. 6.3d), SiCL and SiC soil texture, and mildly alkaline soil pH. The OC contents range from 2.77 to 9.73 g kg^{-1}. The Bd of St soil ranges from 1.28 to 1.72 Mg m^{-3}.

References

Chen ZS, Hseu ZY (1997) Total organic carbon pool in soils of Taiwan. Proc National Sci Council ROC Part B: Life Sci 21:120–127

Foss JE, Moormann FR, Rieger S (1983) Inceptisols. In: Wilding LP, Smeck NE, Hall GF (eds) Pedogenesis and soil taxonomy, Soil orders, vol II. Elsevier, Amsterdam

Hseu ZY, Tsai CC, Guo HY et al (2008) Comparison of Taiwan soil classification system to soil taxonomy and WRB system and its modification toward a common format of Asian soil information systems. In: Proceedings of monsoon asia agro-environmental research consortium (marco) workshop: a new approach to soil information systems for natural resources management in Asian countries. Tsukuba International Congress Center, Tsukuba, Japan, pp 55–69

Soil Survey Staff (1996) Keys to soil taxonomy, 7th edn. SMSS technical monograph No. 19. Pocahontas Press, Inc., Blacksburg

Soil Survey Staff (2006) Keys to soil taxonomy, 10th edn. U.S. Department of Agriculture, Soil Conservation Service, Washington, DC

Tsai CC, Chang YF, Hsu CW et al (2008) Proposed forest soil management divisions (FSMDs) in Taiwan. Ilan Univ J Bioresour 4 (2):99–107 (in Chinese with English abstract)

Tsai CC, Chen ZS, Hseu ZY et al (1998) Representative soils selected from arable and slope soils in Taiwan and their database establishment. Soil Environ 1(1):73–88 (in Chinese with English abstract)

7.1 General Information (Soil Formation Factors and Distribution of Great Group)

Mollisols are characterized with high pH through soil profile and abundant organic matter in the surface soil which meet the definition of mollic epipedon in Soil Taxonomy. Vertisols consist of large amounts of swelling clays like smectites. However, Mollisols and Vertisols are locally found in alluvial or marine terraces from basic or ultrabasic parent rocks in the Coastal Range of eastern Taiwan (Huang et al. 2010; Hseu et al. 2007) (Fig. 7.1).

The Coastal Range comprises two main geological components: (i) island-arc-related Miocene to early Pliocene volcanic andesite rocks, such as lava, pyroclastic breccia, tuff, and epiclastic sandstone and (ii) Pliocene to early Pleistocene sedimentary rocks, such as mudstone, sandstone, and conglomerate, which were deposited in deep-seated fore-arc and inter-arc basins. These geologic units are complex folded or thrust faulted. Three main levels of Holocene marine terraces developed along the eastern Coastal Range are denoted as the first, second, and third levels according to descending altitude (Fig. 7.2). Their formation was via local tectonic uplift. The terraces vary in age on the basis of geomorphic correlation, reflecting their unsynchronized formation through tectonic uplift. The ages of the soils developed on the terrace surfaces were determined via ^{14}C dating of charcoal and by estimations of terrace uplift rates: the soil ages of the first-level terrace are in the range of 7,000–10,000 calibrated (cal.) year, whereas those of the second terrace are younger than 6,000 cal. year, and those of the third-level terrace are younger than 4,000 cal. year. However, Mollisols and Vertisols can be found on the first-level and second-level terraces. The parent materials on the marine terrace supply abundant base cations in association with relatively low rainfall. Additionally, highly activie clay minerals such as smectite and vermiculite are discerned to facilitate the formation of Vertisols and Mollisols (Tsai et al. 2007; Huang et al. 2010).

Histosols are very rare in Taiwan. The landscape of Taiwan is very young and steep, compared to other continental regions in the world. Land reclamation for urban and agricultural development is always imperative in Taiwan, and thus Histosols are hardly found in Taiwan, except on few wetlands behind local convex or lake in the alpine coniferous forest under perhumid climatic conditions along the Central Ridge.

Only one Great Group of Vertisols, Mollisols, and Histosols is recorded in Taiwan, they are Dystruderts, Hapludolls, and Haplohemists respectively.

7.2 Profile Characteristics (Morphology and Micromorphology, Physiochemical Properties, and Mineralogy)

7.2.1 Vertisols from Serpentinites in Eastern Taiwan

Tong-An Mountain derived from serpentinitic rocks is located at Chi-Shan area, Taitung County in the eastern part of Taiwan. Vertisols can be found on the slightly concave of backslope in microrelief. Field morphology of a Vertisol pedon (TA-2) is described as follows (Fig. 7.3). If not otherwise indicated, soil color is given for moist soil.

Z.-S. Chen et al., *The Soils of Taiwan*, World Soils Book Series,
DOI 10.1007/978-94-017-9726-9_7, © Springer Science+Business Media Dordrecht 2015

Aa	0–10 cm	Very dark gray (10YR 3/1) clay loam; strong fine and medium granular structures; cracks with 2 cm in width; firm, slight sticky and slight plastic; many fine and medium roots; clear wavy boundary
AB	10–30 cm	Black (5YR 2.5/1) clay loam; strong medium angular blocky structures; cracks with 1.5 cm in width; very firm, sticky and plastic; few very fine and fine roots; diffuse smooth boundary
Bss1	30–65 cm	Dark reddish brown (5YR 2.5/2) clay; strong medium and coarse angular blocky structures; cracks with 1.5 cm in width; very firm, very sticky and very plastic; few very fine and fine roots; cleary smooth boundary
Bss2	65–94 cm	Very dark brown (7.5YR 2.5/2) clay; strong coarse angular blocky and prismatic structures; cracks with 1.0 cm in width, clear Slickensides on the pedsurfaces; very firm, very sticky and very plastic; few very fine roots; diffuse smooth boundary
Bss3	94–120 cm	Very dark brown (7.5YR 2.5/3) clay; strong coarse angular blocky structures; cracks with 0.5 cm in width, clear slickensides on the pedsurfaces; very firm, very sticky and very plastic; few very fine roots; clear wavy boundary
Bss4	120–145 cm	Dark brown (10YR 3/3) clay; strong coarse angular blocky structures; cracks with 0.5 cm in width, clear slickensides on the pedsurfaces; very firm, very sticky and very plastic; few very fine roots; clear wavy boundary
Bss5	145–165 cm	Dark brown (10YR 3/3) clay; strong coarse subangular blocky structures; cracks with 0.5 cm in width; very firm, very sticky and very plastic; diffuse wavy boundary
BC1	165–200 cm	Dark olive brown (2.5Y 3/3) clay loam; moderate medium subangular blocky structures; firm, sticky and plastic; diffuse wavy boundary
BC2	>200 cm	Dark olive brown (2.5Y 3/3) clay; moderate medium subangular blocky structures; firm, sticky and plastic

Chrysotile, antigorite and lizartide of the serpentine mineral group have similar [Mg$_3$Si$_2$O$_5$(OH)$_4$] composition (Dixon 1989; Caillaud et al. 2006). Antigorite and lizartide are clearly identified under the microscope by different textures (Fig. 7.4). The groundmass serpentine was further identified as antigorite from cross-foliated textures, and as lizardite from hourglass textures. Opaque inclusions of magnetite and chromite stain the parent rock black or brown.

Fig. 7.1 Ultrabasic parent materials (saprolites of serpentine). Copyright by Prof. Hseu

Small amounts of weakly anisotropic light-yellow clay pseudomorphs, and irregular distributed strong-yellow clayey material in voids or on the grain surfaces were interpreted as hypogene neoformed clay minerals in the Vertisol.

Clay content is very high up over 50 % in the Bss horizons of the Vertisol (TA-2 pedon). The OC content decreases with soil depth. The pH is always equal to or higher than 6.0, which is corresponding to high PBS and CEC values. Additionally, the average value of exchangeable Ca/Mg is much lower than 1.0, indicating the characteristics of serpentinitic parent materials (Table 7.1).

The XRD patterns of clay fractions in the Bss1 horizon of TA-2 pedon were illustrated to explain the clay mineralogy of Vertisol (Fig. 7.5). The reflection at 25 °C (room temperature) in the K-saturated clays was present at 0.73 nm, and it could contain chlorite attributed that the peak was reduced when heated to 550 °C. Chlorite was also characterized by the peaks at 1.42, 0.73, 0.475 and 0.35-nm. Chlorite increased with soil depth. Vermiculite was characterized both by the peak at 1.07-nm when the K-saturated clays were heated at 350 and 550 °C and by the peak at 1.42-nm when the Mg-saturated clays were solvated with glycerol. The basal XRD peaks at 1.42-nm with the Mg-saturated clays and collapsed at 1.24-nm with the K-saturated clays heated at 350 and 550 °C are the characteristics of randomly interstratified chlorite-vermiculite (Sawhney 1989; Lee et al. 2003). No depth trend of the interstratified chlorite-vermiculite was found in this pedon, but it was always less abundant in the BC horizons than in their overlying horizons.

Serpentine was also identified in the clay fraction of the pedon by the characteristic XRD peaks at 0.73-nm which persist with K saturation and heating to 550 °C (Istok and Harward 1982). Hseu et al. (2007) indicated that much more

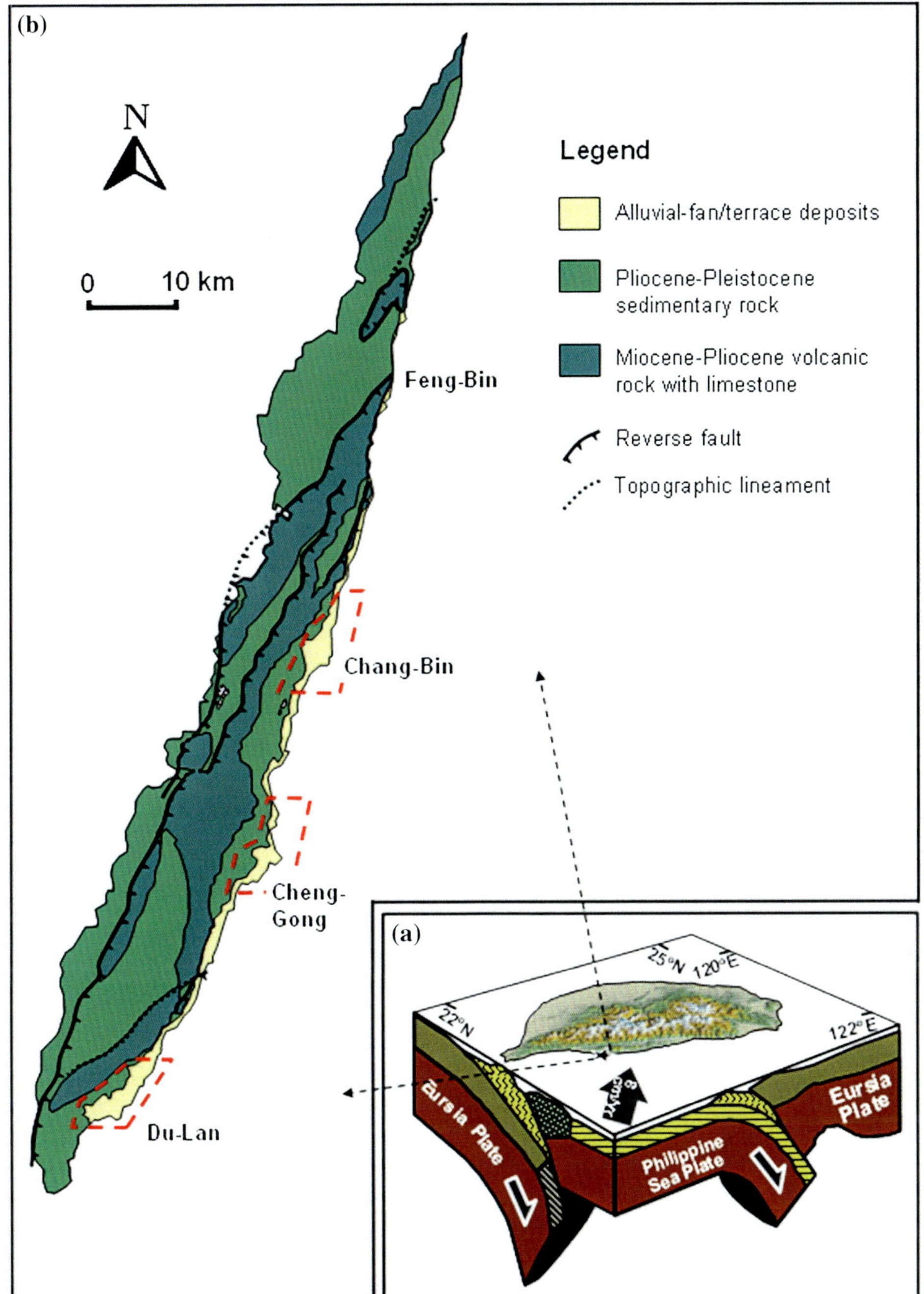

Fig. 7.2 Uplifted marine terraces in Eastern Taiwan: **a** collision between tectonic plates; **b** multiple levels of marine terrace. Copyright by Prof. Hseu

serpentine was found in the younger soils including Vertisols than older soils such as Ultisols on a toposequence, and the serpentine always increased with the soil depth in pedon. This result is in agreement with the report of Graham et al. (1990) which also indicated the serpentine content decreased upward in the profile of a serpentinitic soil in California, USA.

Smectite was identified by its 1.4–1.5-nm spacing after Mg saturation which expands to approximately 1.82-nm after glycerol solvation and which does not completely collapse to 1.0-nm after K saturation and heating to 550 °C for 2 h, suggesting the partial hydroxyl interlayering of smectite. This behavior was already observed in other studies of serpentinite-derived soils (Istok and Harward 1982; Graham et al. 1990). In generally, chlorite, chlorite-vermiculite interstratified minerals, vermiculite, smectite and serpentine are the dominant clay minerals in the pedon.

Fig. 7.3 Field morphology of TA-2 pedon on the flat landscape of serpentine in eastern Taiwan (*left*) which clear slickensides are on the pedsurfaces (*right*). Copyright by Prof. Hseu

7.2.2   Vertisol and Mollisols on the Marine Terraces of Eastern Taiwan

Three different orders of soils have developed on marine terraces along the eastern coast of Taiwan, and their distribution relates to elevations of uplifted terraces (Tsai et al. 2007). Huang et al. (2010) selected 13 pedons along three transects (Chang-Bin, Cheng-Gong, and Du-Lan) in the central and southern part of the eastern coast of Taiwan to explore the degree of development of these soils (Fig. 7.2). Entisols were always found on the youngest terrace (3rd level), Mollisols typified the intermediate terrace (2nd level), and Vertisols predominated on the oldest terrace (1st level). Hapluderts are distributed on the first-level terrace along the Chang-Bin and Cheng-Gong transects. However, Hapludolls are located on the first-level terrace along the Du-Lan transect, and are the main soil groups on the second-level terraces of all three transects. Hapludolls are also distributed on the third-level terrace of the Cheng-Gong transect, despite Udipsamments being predominant on all transects of the third-level terraces.

The soils on the terraces of the third, second, and first levels tend to show increasing development with increasing altitude. Consequently, the soils form a sequence matching terrace levels from young (3rd and intermediate 2nd terraces) to old (1st terrace): Udipsamments, Hapludolls, and Hapluderts in *Soil Taxonomy* (Soil Survey Staff 2010). This finding of soil classification supports the pathway of soil development from Entisols, to Mollisols, and to Vertisols through time. In addition, this pattern agrees well with the concept of a "post-incisive" type soil chronosequence (Vreeken 1975), with soils of different ages ending their development synchronously. Field morphology of a Vertisol (CG-1 pedon) is described as follows (Fig. 7.6). If not otherwise indicated, soil color is given for moist soil.

A	0–10 cm	Very dark brown (10YR 2/2) clay; strong very coarse and coarse angular blocky structures; cracks with 2 cm in width, some slickensides; very firm, very sticky and very plastic; many fine and medium roots; clear wavy boundary
Bss1	10–40 cm	Very dark gray (10YR 3/1) clay; strong very coarse and coarse angular blocky structures; cracks with 2 cm in width, some slickensides; very firm, very sticky and very plastic; some fine roots; smooth wavy boundary
Bss2	40–65 cm	Very dark gray (2.5YR 3/1) clay; strong medium angular blocky structures; cracks with 1 cm in width; very firm, very sticky and very plastic; many fine roots; smooth wavy boundary
R	>65 cm	Very dark gray (2.5Y 3/1) clay; 65 % stones; moderate medium subangular blocky structures; firm, slight sticky and slight plastic

Field morphology of a Mollisol pedon (CG-3) is described as follows (Fig. 7.6). If not otherwise indicated, soil color is given for moist soil.

A	0–15 cm	Grayish brown (2.5Y 5/2) loam; strong medium granular structures; firm, slightly sticky and slightly plastic; many fine and medium roots; diffuse smooth boundary
Bw1	15–30 cm	Dark olive brown (2.5Y 3/3) clay loam; strong medium and coarse angular blocky structures; firm, sticky and plastic; many fine roots; diffuse smooth boundary
Bw2	30–50 cm	Dark yellowish brown (10YR 4/6) clay loam; strong medium angular blocky structures; firm, slightly sticky and plastic; many very fine roots; diffuse smooth boundary
C1	50–70 cm	Dark yellowish brown (10YR 3/6) sandy clay loam; moderate fine and medium subangular blocky structures; firm, slightly sticky and slightly plastic; some very fine roots; diffuse smooth boundary
C2	70–110 cm	Black (2.5YR 2.5/1) sandy loam; moderate fine and medium subangular blocky structures; firm, slightly sticky and slightly plastic; diffuse smooth boundary
C3	>110 cm	Dark olive brown (2.5Y 3/3) sandy; weak fine subangular blocky structures; friable

Surface cracks are very clear in the Vertisols. In the observation of the thin sections, stress coatings are also the main pedofeatures of the Vertisol on the first- and second-level terraces (Fig. 7.7); however, the degree of stressing on the second-level terrace is lower than that on the first-level terrace. Regarding with the swelling-shrinkage pedoturbation by clay materials, the soils on the second-level terrace also have evidence of plasma modification like the Vertisols on the first-level terrace. However, they differ from the typical Vertisols in failing to crack sufficiently in having low amounts of swelling-type clay to meet the criteria in slickenside for Vertisols. Additionally, the unique micromorphological characteristics of the soils on the second-level terrace is characterized by gefuric C/F related distribution which the b-fabric is undifferentiated, and thus the altered coarse fragments are surrounded by homogeneous fine materials.

Bulk densities in all horizons are in the range between 1.3 and 2.1 Mg/m^3, and the Vertisol appears to be higher than the Mollisol (Table 7.2). The clay contents correspond to field observations regarding the soil structure, stickiness, and plasticity. Namely, the Vertisol on the first-level terrace has composes higher clay than the Mollisol on the second-level terrace. Unlike on the other terrace levels, the sand fractions of CG-1 pedon on the first-level terrace are dominated by fine and very fine sand, suggesting that finer soil particles of

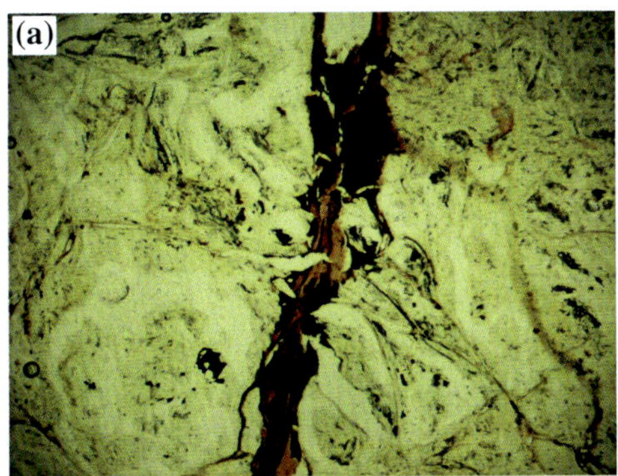
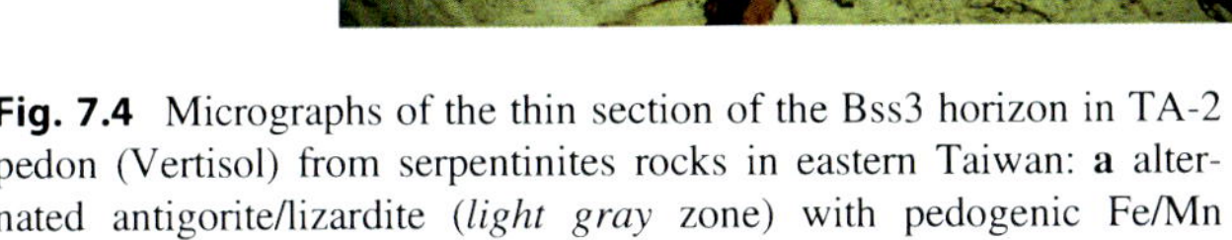

Fig. 7.4 Micrographs of the thin section of the Bss3 horizon in TA-2 pedon (Vertisol) from serpentinites rocks in eastern Taiwan: **a** alternated antigorite/lizardite (*light gray* zone) with pedogenic Fe/Mn oxides in the void under PPL; length of frame is 1.25 mm; **b** the same to (**a**) but with XPL. Copyright by Prof. Hseu

Table 7.1 Selected physical and chemical properties of TA-2 pedon

Horizon	Depth (cm)	Sand (%)	Silt (%)	Clay (%)	OC[a] (%)	pH	PBS[b] (%)	CEC[c] (cmol/kg)	Fe$_d$[d] (g/kg)	Ca/Mg[e]
Aa	0–10	32	32	36	2.9	6.5	35	32.0	14.0	0.5
AB	10–30	41	26	33	2.2	6.4	60	27.4	15.3	0.3
Bss1	30–65	27	22	52	1.5	6.0	66	36.3	14.6	0.2
Bss2	65–94	28	15	58	1.2	6.0	55	34.6	19.7	0.2
Bss3	94–120	25	20	56	1.0	6.0	59	40.3	28.8	0.2
Bss4	120–145	34	24	53	1.0	6.2	55	35.5	12.2	0.2
Bss5	145–165	24	26	50	0.8	6.0	57	37.8	16.3	0.2
BC1	165–200	42	21	37	0.7	6.1	61	51.1	13.4	0.2
BC2	>200	34	25	41	0.8	6.0	62	36.3	14.2	0.2

[a] Organic carbon
[b] Percentage of base saturation
[c] Cation exchange capacity
[d] DCB-extractable Fe
[e] Ca and Mg are exchangeable form
Copyright by Prof. Hseu

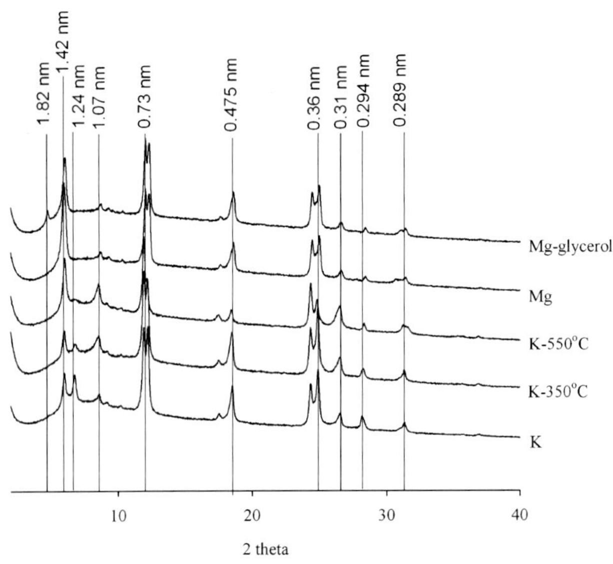

Fig. 7.5 X-ray diffraction pattern of clay fraction of the Bss1 horizon in Vertisol from serpentinites in Eastern Taiwan. *K* K-saturated treatment; *Mg* Mg-saturated treatment. Copyright by Prof. Hseu

Vertisols occur on the first-level terrace than those of Mollisols on the second-level terrace in eastern coastal Taiwan because of difference in time span of weathering.

The pH values of the Vertisol and Mollisol on the marine terraces of eastern Taiwan are mainly slightly acidic to neutral (Table 7.3). There is abundant OC in their surface horizons; in comparison with the global OC figures, the soils were low in OC. This could be because of rapid decomposition of the biomass under the prevalent hyperthermic temperature regime and udic conditions and poor vegetation cover on the marine terraces. Therefore, organic carbon is less a function of bulk density than iron oxide and clay in this study. Small quantities of pedogenic oxides, including iron (Fe$_d$ and Fe$_o$) and aluminum (Al$_d$ and Al$_o$) oxides, were found in the soils. Generally, the soil CECs are high, with values exceeding 15 cmol/kg, and the base saturations reach 30 % or more throughout all soils. The increase of CEC with soil age depends on clay content, OC accumulation, and the presence of expandable clay minerals such as smectite, as identified by Tsai et al. (2007) for the Entisols, Mollisols, and Vertisols on the marine terraces of eastern Taiwan.

Fig. 7.6 Vertisols (*left*) and Mollisols (*right*) on marine terrace in Eastern Taiwan. Copyright by Prof. Hseu

Fig. 7.7 Surface crack in the field (*left*) and micromorphological characterisitcs of stress coating (*right*, XPL) in the Vertisol (CG-1 pedon) on marine terrace in Eastern Taiwan. Copyright by Prof. Hseu

Table 7.2 Bulk density and particle size distribution of CG-1 and CG-3 pedons on the marine terraces in Eastern Taiwan

Horizon	Depth	BD[+a]	Sand	Silt	Clay	Sand fractions[b]				
						VC	C	M	F	VF
CG-1 pedon (Vertisol)										
A	0–10	1.6	12	52	35	0.3	0.2	0.4	3.2	7.8
Bss1	10–40	2.1	16	43	42	0.1	0.3	1.0	4.8	9.9
Bss2	40–65	1.8	15	40	46	0.1	0.5	1.5	5.7	7.0
CG-3 pedon (Mollisol)										
A	0–15	1.5	38	35	27	0.8	1.9	6.3	17	12
Bw1	15–30	1.9	31	40	29	0.3	0.9	3.4	10	17
Bw2	30–50	1.9	44	27	30	0.8	2.4	14	22	5.3
C1	50–70	1.8	71	9.0	21	1.1	5.1	23	35	6.1
C2	70–110	2.0	79	6.0	14	1.7	12	34	27	4.5
C3	110–130	1.9	83	7.0	10	4.3	16	48	10	3.8
CR	>110	1.3	87	5.0	8.0	11	41	20	7.1	7.3

[a] BD which definition is bulk density
[b] VC very coarse; C coarse; M medium; F fine; VF very fine
Copyright by Huang et al. (2010)

Table 7.3 Selected chemical properties of CG-1 and CG-3 pedons on the marine terraces in Eastern Taiwan

Horizon	Depth (cm)	pH	OC[+] (g/kg)	Fe_d^a (g/kg)	Al_d (g/kg)	Fe_o (g/kg)	Al_o (g/kg)	CEC[b] (cmol/kg)	BS[c] (%)
CG-1 pedon (1st terrace)									
A	0–10	5.4	21	2.11	0.38	1.86	0.08	30	37
Bss1	10–40	6.7	8.0	2.33	0.59	1.15	0.12	36	41
Bss2	40–65	6.9	6.1	2.33	0.59	1.15	0.12	44	42
CG-3 pedon (2nd terrace)									
A	0–15	5.9	26	2.80	0.47	2.51	0.08	25	37
Bw1	15–30	6.2	13	4.32	0.98	1.66	0.11	26	40
Bw2	30–50	6.9	13	6.18	1.69	0.55	0.12	23	38
C1	50–70	7.1	5.7	5.43	0.93	0.63	0.08	23	73
C2	70–110	7.0	3.0	2.30	0.37	0.70	0.08	16	44
C3	110–130	7.2	3.0	1.22	0.25	0.61	0.08	18	100

[a] DCB-extractable metals (Fe_d and Al_d); oxalate-extractable metals (Fe_o and Al_o)
[b] Cation exchange capacity
[c] Base saturation
Copyright by Huang et al. (2010)

7.2.3   Histosols in the Central Ridge

7.2.3.1 Chinanshan Pedon

There are some lakes in the closed depressions on high mountain areas along the Central Ridge in association with dense coniferous forests. Additionally, the climate is generally characterized by cool and humid conditions in these ecosystem, so that a large amount of organic matter from plant debris was accumulated in the lake to form Histosols. This pedon is located in I-Lan, northeastern Taiwan (N24° 34′ 45″, E121° 23′ 46″), which elevation is 1,800 m and the annual rainfall is about 3,500 mm. The average air temperature is 13 °C and soil water regime is perudic. The geological stratum mainly consists of shale. The dominant vegetation is *Rhododendron formosanum*.

Field morphology of the Chinanshan pedon is described as follows. If not otherwise indicated, soil color is given for moist soil.

Oi	0–10 cm	Black (5YR 2.5/1) slightly decomposed organic materials; rubber fiber content is about 10 % of the volume; weak very fine and fine granular structures; many very fine and fine roots; clear wavy boundary
Oe	10–20 cm	Black (2.5YR 2.5/2) intermediate decomposed organic materials; rubber fiber content is about 25–30 % of the volume; weak very fine and fine granular structures; many fine and medium roots; clear wavy boundary
Oa	20–32 cm	Black (7.5YR 2/0) highly decomposed organic materials; rubber fiber content is about 50 % of the volume; weak very fine and fine granular structures; many medium roots; abrupt wavy boundary

The soil pH was very low because of the great contribution of plant residue to acidity. These residues also trapped a clear amount of Al and H ions, so that the pH measured by $CaCl_2$ was below 2.5 which was much lower than the pH by water (Table 7.4). The OC content exceeded 30 % in all horizons, while exchangeable bases and BS value was very low. These characteristics represent typical Histosols worldwide.

7.2.3.2 Gutzelun Pedon

This pedon is located in Pingtung which belongs to the southern tip of the Central Ridge (N22° 25′ 08″, E120° 47′ 16″) with the elevation of about 1,630 m above the sea level. The major stratum is slate and schist. The annual rainfall in this area is around 5,000 mm by the great impact of humid clouds from southern Pacific Ocean, so that hardwood and coniferous trees are very dense in these tropical forests with abundant ground cover vegetation of herb and grass. In general, the landscape is very steep (>30 %), but some microrelief is a closed depression like the point of Gutzelun pedon (Fig. 7.8). Large amounts of plant debris have been long-term accumulated on these concave ground, which are mainly from *Rhododendron rubropilosum* Hayata var. taiwanalpinum. Therefore, a typical Histosol is formed. However, the plant debris is not decomposed well and cell residues are still discerned by light microscopy (Fig. 7.9).

Field morphology of the Gutzelun pedon is described as follows. If not otherwise indicated, soil color is given for moist soil.

Table 7.4 Bulk density and chemical properties of Chinanshan pedon

Horizon	Depth (cm)	pH		OC (%)	Exchangeable base				CEC (cmol/kg)	BS (%)
		H_2O	$CaCl_2$		K (cmol/kg)	Na (cmol/kg)	Ca (cmol/kg)	Mg (cmol/kg)		
Oi	0–10	3.2	2.3	31.9	1.69	7.11	0.29	0.09	97.4	9.4
Oe	10–20	3.4	2.3	31.7	1.5	0.99	0.3	0.07	87.1	3.3
Oa	20–32	3.3	2.2	31.3	1.64	5.74	0.29	0.1	108	7

Copyright by Huang et al. (2010)

Oe	0–15 cm	Black (2.5Y 2.5/1) intermediate decomposed organic materials; rubber fiber content is about 30 % of the volume; weak very fine and fine granular structures; many very fine and fine roots; clear wavy boundary
Oi	15–30 cm	Black (2.55Y 2.5/1) slightly decomposed organic materials; rubber fiber content is about 50 % of the volume; weak very fine and fine granular structures; many very fine and fine roots; abrupt wavy boundary

Like Chinanshan pedon from northern Taiwan, the soil pH was still very low in Gutzelun pedon from southern Taiwan (Table 7.5). However, Gutzelun pedon has much more OC content than Chinanshan pedon, because the plant residue in Gutzelun pedon was mainly from woody species in association with higher rainfall compared to Chinanshan pedon. However, exchangeable bases and BS value was very low in this pedon.

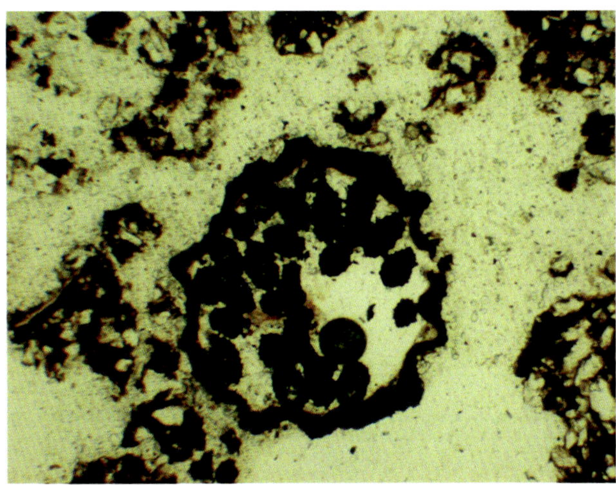

Fig. 7.9 Cell residues of plant tissues of Gutzelun pedon. Copyright by Prof. Hseu

Fig. 7.8 Morphological characteristics of Gotzelun pedon. Copyright by Prof. Hseu

7.3 Formation and Genesis

The distribution of soil type may be closely related to the history of landscape evolution at a site and across a range of landforms (McFadden and Knuepfer 1990), such as the Vertisols and mollisols on the marine terraces of eastern Taiwan (Tsai et al. 2007; Huang et al. 2010). The mountain ranges of Taiwan are the result of arc-continent collision, as noted earlier. The Coastal Range, one of the youngest elements of the Taiwan orogenic belt, is the only place in eastern Taiwan that is fringed with multiple levels of marine terraces arising from the collision (Fig. 7.2). Soils begin to develop on marine terraces when the land surface is uplifted above sea level. Consequently, the uplift rates of marine terraces can be examined by comparing the degree of soil development on a given terrace level along the coast.

Table 7.5 Bulk density and chemical properties of Gutzelun pedon

Horizon	Depth (cm)	Bd (Mg/m^3)	pH		OC (%)	Exchangeable base				CEC (cmol/kg)	BS (%)
			H$_2$O	CaCl$_2$		K (cmol/kg)	Na (cmol/kg)	Ca (cmol/kg)	Mg (cmol/kg)		
Oe	0–15	0.12	3.37	2.35	91.5	0.40	0.08	0.53	0.17	120	2.33
Oi	15–30	0.14	3.22	2.50	55.4	0.71	0.20	1.05	0.25	284	3.54

According to the U.S. Soil Taxonomy (Soil Survey Staff 2010), three soil orders can be identified on different levels of marine terraces in eastern Taiwan (Entisols, Mollisols, and Vertisols). However, Hapluderts are distributed on the first-level terrace. Hapludolls and are the main soil groups. These soils display a "post-incisive" type of soil chronosequence, suggesting a developmental sequence from Hapludolls to Hapluderts. This sequence is common to soils on marine terraces in a range of different climates. However, the advanced degree of development of the soils on the older terraces indicates very rapid soil development despite their Holocene ages in Taiwan.

7.4 Land Use and Management

Paddy rice is the major staple crop in Taiwan, with less extent of corn. Consequently, in the marine terraces of eastern Taiwan, Vertisols and Mollisols are very fertile for these crop production because of their abundant base cation concentrations corresponding to their features of high CEC and neutral pH regimes. After the Second World War, these areas had been developed for paddy rice production. However, typhoon event frequently occurs annually in these areas, and thus the paddy soils with Vertisols and Mollisols are not so productive in rice. Additionally, the international rice market has been liberalized since the 2000s in Taiwan, indicating the total yield of rice became lower than before controlled by government policy. These soils have been used for other crops or for non-agricultural application. However, changing the land use from paddy soils into non-waterlogged cropping has some problems in initial soil properties such as poor drainage and impeded root growth by the subsurface compacted layers for upland crops particularly in the Vertisols with very high clay content. Another problem raises from this changing of land use in the Vertisols and Mollisols. Paddy soil management significantly increases the organic carbon in the surface soil horizons, because the upper parts of a typical paddy soil is composed of a plow layer and a plowpan (Cheng et al. 2009). If these soils are stopped being submerged, the function of soil in carbon sequestration will be reduced. Therefore, the land owners should be encouraged to increase vegetative covers for these soils not only for carbon sequestration but also for reduction of wind and water erosion in the coastal areas with Vertisols and mollisols.

Regarding Histosols in Taiwan, they are always located in the preserved areas of high mountain forests. These soils have not been disturbed by any human activity, whose major function in ecology is to maintain nature sustainability.

References

Caillaud J, Proust D, Righi D (2006) Weathering sequences of rock-forming minerals in a serpentinite: influence of microsystems on clay mineralogy. Clays Clay Miner 54:87–100

Cheng YQ, Yang LZ, Cao ZH, Ci E, Yin S (2009) Chronosequential changes of selected pedogenic properties in paddy soils as compared with non-paddy soils. Geoderma 151:31–41

Dixon JB (1989) Kaolin and serpentine group minerals. In: Dixon JB, Weed SB (eds) Minerals in soil environments, 2nd edn. SSSA Book Series No. 1. SSSA, Madision, pp 467–526

Graham RC, Diallo MM, Lund LJ (1990) Soils and mineral weathering on phyllite colluviun and serpentinite in northwestern California. Soil Sci Soc Am J 54:1682–1690

Hseu ZYH, Tsai HC, Hsi, Chen YC (2007) Weathering sequences of clay minerals in soils along a serpentinitic toposequence. Clays Clay Miner 55:389–401

Huang WS, Tsaia H, Tsai CC, Hseu ZY, Chen ZS (2010) Subtropical developmental soil sequence of Holocene marine terraces in Eastern Taiwan. Soil Sci Soc Am J 74:1271–1283

Istok JD, Harward ME (1982) Influence of soil moisture on smectite formation in soils derived from serpentinite. Soil Sci Soc Am J 46:1106–1108

Lee BD, Sears SK, Graham RC, Amrhein C, Vali H (2003) Secondary mineral genesis from chlorite and serpentine in an ultramafic soil toposequence. Soil Sci Soc Am J 67:1309–1317

McFadden LD, Knuepfer PLK (1990) Soil geomorphology: the linkage of pedology and surficial process. Geomorphology 3:197–205

Sawhney BL (1989) Interstratification in layer silicates. In: Dixon JB, Weed SB (eds) Minerals in soil environments, 2nd edn. SSSA Book Series No. 1. SSSA, Madision, pp 789–828

Soil Survey Staff (2010) Keys to soil taxonomy. Natural resources conversation services, 12th edn. United States Department of Agriculture, Washington, DC

Tsai CC, Tsai H, Hseu ZY, Chen ZS (2007) Soil pedogenesis along a chronosequence on marine terraces in eastern Taiwan. Catena 71:394–405

Vreeken WJ (1975) Principal kinds of chronosequences and their significance in soil history. J Soil Sci 26:378–394

Spodosols

8

8.1 General Information (Soil Formation Factors and Distribution of Great Group)

Spodosols are common in forests of Nordic countries, North America, and northern parts of Russia, which occur mainly in cool humid climates under forest vegetation in medium textured to coarse soil materials, where conditions favor the development of an organic surface layer (McKeague et al. 1983). Below the organic layer is an ash-gray, weathered eluvial horizon (E), which contains less base cations, Al, and Fe than the parent material and is enriched in residual Si. This horizon is underlain by a dark colored (reddish-brownish, dark brown, black) illuvial horizon (Bhs) enriched Al, Fe, and organic materials. Acidic coarse parent materials and litter from the coniferous trees are conductive for podzolization of these soils (Lyford 1952; De Coninck 1980).

Coarse textured Spodosols are not commonly found in Taiwan. The Spodosols of Taiwan are distributed mostly along the Central Ridge, which is located at an elevation of higher than 1,800 m, under cool and humid climate with high precipitation (>3,000 mm/yr), and is covered dominantly with coniferous trees (Fig. 8.1). Soils with various degrees of podzolization have been performed in these coniferous ecosystems (Li et al. 1998; Hseu et al. 1999, 2004; Lin et al. 2002; Liu and Chen 1998, 2004). Entisols, Inceptisols, and Ultisols are preferably found under this environment. A significant white eluvial ablic horizon (E) and dark reddish illuvial B horizon coexist in general in podzolic soils of Taiwan, especially in the mountains of the Central Ridge (Fig. 8.2), but some of these B horizon do not completely meet the criteria of the morphological characteristics and chemical properties defined for a spodic horizon in Soil Taxonomy (Soil Survey Staff 2006). The Spodosols generally form at the flatter summit and backslope in microrelief (Li et al. 1998). Li et al. (1998) and Lin et al. (2002) indicated that both spodic material formation and clay translocation occurred in the Bhs horizons of loamy Spodosols in central Taiwan.

Spodosols of Taiwan significantly differ from those of North American and European countries in terms of soil texture and precipitation. Taiwan's Spodosols are developed in fine-textured materials such as shale and slate, and also in transported materials under heavy annual rainfall, which intensifies the lessivage and leaching processes. Clay contents in the Spodosol or podzolic soils are generally more than 30 %, and both podzolization and clay illuviation occur in these soils of Taiwan (Chen et al. 1989, 1995; Li et al. 1998; Hseu et al. 1999; Lin et al. 2002). Van Ranst et al. (1997) also found cool tropical Podzols in Rwanda, Central Africa, which was characterized by a cool isomesic soil temperature regime that prevails at high elevations in areas with the parent materials of quartzites and micaceous sandstones along steep slopes. In particular, understanding the transition between Ultisols and Spodosols by illuviation of clay and spodic materials in the subalpine forests of central Taiwan is important for establishing the mechanism of podzolization.

Podzolization is a formation process of spodic materials defined as accumulated complexes of organic acids with Al and Fe in the subsurface horizons. Not only Spodosols but also other types of soils form spodic materials. However, the key requirement for Spodosol in Soil Taxonomy is spodic horizon whose required characteristics are, it must have 85 % or more spodic materials in a layer 2.5 cm or more thick that is not part of any Ap horizon. Therefore, four types of podzolic soils can be extensively found in Taiwan attributed to various degrees of podzolization. These podzolic soils are Spodosols, Ultisols, Inceptisols without placic horizon, and Inceptisols with placic horizon. In this chapter, two representative types of Spodosols in Taiwan are introduced as Typic Haplorthods and Ultic Haplorthods, while the other podzolic soils are discussed in the other chapters.

Z.-S. Chen et al., *The Soils of Taiwan*, World Soils Book Series,
DOI 10.1007/978-94-017-9726-9_8, © Springer Science+Business Media Dordrecht 2015

Fig. 8.1 Vegetation under perudic moisture regime conditions of Taiwan. Copyright by Profs. Chen, Hseu, and Tsai

Fig. 8.2 Profile morphology of typical Spodosols in Taiwan. Copyright by Prof. Chen

8.2 Profile Characteristics

8.2.1 Typic Haplorthods

Compared to southern Taiwan, it is easier to form Spodosols in the alpine and subalpine forests in northern and middle Taiwan because these regions directly face the northeast monsoon from China to produce cooler and more humid climate than other regions of Taiwan. Typic Haplorthods from central Taiwan are typical Spodosols.

8.2.1.1 JL Pedon

The profile described below (JL pedon) is situated in Jen-Lun forestry area (23° 42′ N, 120° 56′ E) in Nantou County, central Taiwan and at an elevation of 2,470 m above sea level (Fig. 8.3). The mean annual air temperature, annual precipitation, and average relative humidity are 12.7 °C, >3,000 mm, and 72 %, respectively. Seventy-six percent of precipitation falls from May to September. The soil moisture regime and soil temperature regime are udic and mesic, respectively. The parent materials consist of shale and slate. The native vegetation is dominantly Chinese hemlock (*Tsuga chinensis* Pritz), but after cutting and burning of the native forest, the secondary forests are dominantly Taiwan red pine (*Pinus taiwanensis* Hayata), Yushan cane [*Yusania niitakayamensis* (Hay.) Keng F.], and alpine silver grass (*Miscanthus tmnsmorrisonensis* Hayata). The pedon has a cover of secondary forest consisting mainly of Taiwan red cypress

Fig. 8.3 Field morphology of the JL pedon in Nantou County, central Taiwan. Copyright by Prof. Chen

(*Chamaecyparis formosensis* Matsum) and peacock pine [*Cryptomeria japonica* (L.f.) D. Don] species planted by the Taiwan Forestry Bureau in the 1950s.

Field morphology of the JL pedon is described as follows. If not otherwise indicated, soil color is given for moist soil.

O/A	0–3 cm	Grayish brown (10YR 5/2) silt loam; moderate very fine and fine granular structures; friable; many very fine and fine roots; clear smooth boundary
E	3–20 cm	Light gray (10YR 7/2) silt loam; massive structure; slight sticky and slight plastic; common very fine and fine roots; abrupt smooth boundary
Bh	20–32 cm	Strong brown (7.5YR 4/6) clay; strong very fine angular blocky and moderate very fine and fine granular structures; firm, slight sticky and slight plastic; common very fine and fine roots; diffuse smooth boundary
Bw	32–50 cm	Strong brown (7.5YR 5/8) silty clay; strong very fine and fine angular blocky and moderate very fine and fine granular structures; firm, slight sticky and slight plastic; common very fine and fine roots; gradual smooth boundary

(continued)

(continued)

| 2Bw | 50–70 cm | Strong brown (7.5YR 5/8) silty clay; strong fine and medium angular blocky structures; slight sticky and slight plastic; common very fine and fine roots; gradual smooth boundary |
| 2C | 70–90 cm | Brownwish yellow (10YR 6/8) clay loam; strong fine angular blocky structures; firm, slight sticky and slight plastic; common very fine and fine roots; abrupt wavy boundary |

This profile has a dark organic/mineral (O/A) surface, a well-developed albic (E) horizon, and spodic (Bh) horizon. These spodic materials qualify the pedon as Spodosol based on the color criteria of spodic materials as defined in Soil Survey Staff (2006).

In terms of micromorphology, microaggregates are indicated as dark pellets primarily composed of irregular organo-Fe (or -Al) complexes and illuvial clays (Fig. 8.4). Isotropic illuvial clays are accumulated or mixed with organo-Fe complexes along the irregular voids in the Bh horizons. The clay content of the spodic horizon is quite high. The illuvial clay can be found in the irregular voids. These micromorphological characteristics indicate that the illuviation of clay has occurred in the Bh horizons. However, the podzolization process partially inhibited the orientation of this illuvial clay in the voids or on the ped surface.

The clay content and chemical properties of the pedon are listed in Table 8.1. Because the parent materials are shale and slate, the pedon has unusually fine textures in the subsurface horizons, especially in the Bh horizon, which has high clay content. In general, the texture of the pedon is finer

Fig. 8.4 Photomicrograph of a thin section in the Bh horizon (20–32 cm) the JL pedon in Nantou County, central Taiwan. The microaggregates are indicated as *dark pellets* primarily composed of irregular organo-Fe complexes and illuvial clay (PPL). The width of the photo frame is 0.625 mm. Copyright by Prof. Hseu

Table 8.1 The physical and chemical properties of the JL pedon in Nantou County, central Taiwan

Horizon	Clay	pH	OC^a (g/kg)	CEC^b (cmol/kg)	Exchangeable bases				BS^c (%)
					K (cmol/kg)	Na (cmol/kg)	Ca (cmol/kg)	Mg (cmol/kg)	
O/A	18	3.5	102	31	0.24	0.04	0.02	0.36	2
E	25	3.7	16	15	0.09	0.02	ND^d	0.23	2
Bh	58	4.4	44	29	0.32	0.05	ND	0.37	3
Bw	47	4.7	17	14	0.11	0.02	ND	0.17	2
2Bw	42	4.9	10	10	0.18	0.02	ND	0.39	6
2C	38	5.0	7	8	0.15	0.02	0.01	0.43	8

[a] Organic carbon
[b] Cation exchange capacity
[c] Base saturation
[d] Not detectable
Copyright by Prof. Hseu

than those in temperate regions (McKeague et al. 1983). The pH value ranges from 3.5 to 5.0, which is lower than the criterion defined for spodic materials (pH $\leq$ 5.9). The O/A and E horizons tend to have lower pH than B horizons. Organic C content is less in E horizons and greater in B horizons, especially in Bh horizon. This indicates that organic C has eluviated from albic horizons and accumulated in the spodic horizon. Very low base saturation (mostly <4 %) in the pedon is due to strong leaching processes. Cation exchange capacity is influenced primarily by the amount of organic matter, amorphous material, and clay content in the soils. Because these materials are eluviated from upper horizons, CEC values are higher in spodic horizon than in E horizon.

The concentrations and ratios of the forms of extractable Fe and Al determined by DCB and ammonium oxalate extraction methods are shown in Table 8.2. Extractable Fe and Al contents were significantly higher in Bh horizon than in E horizon for Spodosol. It indicates illuviation and accumulation of Fe and Al from E horizons into the spodic horizon. The maximum contents of extractable Fe as measured by the extraction methods were located in the upper part of the Bh horizon. The maximum content of extractable

Al was found near the middle part of the B horizons. The value of DCB-extractable Fe in the B horizons drastically increased in comparison to that in the respective E horizon. Clay, Fe, and organic C maxima coincided in the same Bh horizon of the pedon but Al maxima differed slightly. This supports the illuviation of clay, organic C, and Fe in the Bh horizon. The dominant color of the B horizons is 7.5YR 4/6. According to Soil Survey staff (2006), the chemical characteristics of the B horizon must indicate that enough spodic materials in the subhorizons have been accumulated to classify them as Spodosols. The ODOE values of B horizons are >0.25, and each value is at least two times as high as the ODOE value for an overlying eluvial horizon. Furthermore, in the B horizons, total Al plus half the Fe content extracted by ammonium oxalate is greater than 0.5, and each value is also at least two times as high as the value for an overlying eluvial horizon. These increased ODOE values and $Al_o + 1/2Fe_o$ percentages in the upper part of the B horizons indicate that the accumulation and translocation of spodic materials such as organic matter, Fe, and Al have occurred in illuvial horizons of the Spodosols.

8.2.1.2 TS Pedon

The other pedon (TS pedon) of Typic Haplorthods introduced in this chapter is located in Tzu-Shan area of A-Li Mountain in subalpine region of Central Taiwan. The elevation is about 2,300 m. The vegetation types are dominated by Taiwan red cypress (*Chamaecyparis formosensis*) and Chinese hemlock (*Tsuga chinensis*), with minor species of pine (*Pinus armandii*) and Wheel Stamen tree (*Trochdendron aralioides*). The climate includes a mean annual air temperate of 10.8 °C and a mean annual precipitation of approximately 4,000 mm, most of it falls from May to September. Soil moisture regime is udic and the soil temperature regime is mesic. The soil has developed from marine argillaceous sediments and interstratified sandstone and shale of the late Miocene to early Pliocene age.

Table 8.2 The contents and ratios of Fe and Al in the JL pedon in Nantou County, central Taiwan

Horizon	Dithionite		Oxalate		$Al_o + 1/2Fe_o$	$ODOE^a$
	Fe_d (g/kg)	Al_d (g/kg)	Fe_o (g/kg)	Al_o (g/kg)		
O/A	14.8	3.0	5.5	2.4	0.52	0.72
E	11.6	2.0	3.2	1.6	0.32	0.16
Bh	81.8	19.5	26.9	9.0	2.24	0.98
Bw	54.3	15.2	21.7	7.9	1.88	0.59
2Bw	42.7	11.4	17.4	6.0	1.47	0.41
2C	34.4	9.0	14.3	4.9	1.20	0.24

[a] Optical density of the oxalate extraction
Copyright by Prof. Hseu

Field morphology of the TS pedon is described as follows. If not otherwise indicated, soil color is given for moist soil.

Oi	0–5 cm	Black (7.5YR 2.5/1) moderate very fine and fine granular structures; friable; many very fine and fine roots; gradual wavy boundary
A	5–15 cm	Black (10YR 2/1) sand; strong very fine angular blocky and moderate very fine and fine granular structures; firm, slight sticky and slight plastic; common very fine and fine roots; diffuse smooth boundary
AE	15–20 cm	Gray (7.5YR 5/1) loamy sand; structureless; very friable; few very fine roots; diffuse smooth boundary
E	20–30 cm	Gray (7.5YR 6/1) loamy sand; structureless; very friable; few very fine roots; gradual smooth boundary
Bhs	30–43 cm	Strong brown (7.5YR 4/6) loamy sand; moderate very fine angular blocky structure; friable; few very fine roots; diffuse smooth boundary
Bw	43–48 cm	Brownish yellow (10YR 6/8) loamy sand; moderate very fine augular blocky structure; firm; few very fine roots; clear smooth boundary
C	48.70 cm	Yellowish brown (10YR 5/8) loamy sand; moderate very fine and fine augular blocky structures; friable; few very fine roots

The pedon has a dark organic/mineral (O & A) surface, a well-developed albic (E) horizon, and spodic (Bhs) horizon, which is classified as Spodosols based on the criteria in Soil Taxonomy (Soil Survey Staff 2006). The color of the Albic E horizon is gray (7.5YR 6/1). The soil colors in all of the subsurface horizons with different degrees of spodic material accumulation are close to 7.5YR 5/6, which is defined in Soil Taxonomy as one of soil morphological criteria for the spodic horizon (Bhs) (Soil Survey Staff 2006). The changes in soil color and soil texture throughout the soil profiles are distinct. The color were gray or light gray, with a soil chroma 1, in eluvial horizon, but changed to strong brown or brownish yellow color in the subsurface horizons.

The physical properties of TS pedon are presented in Table 8.3. This pedon without clay accumulation is characterized by many voids in the spodic horizon. In general, the size of soil texture is finer than those of temperate regions (McKeaque et al. 1983; Sommer et al. 2001). The sandy and loamy Spodosols prevailed in the area surrounding TS pedon due to an increase in clay content. This phenomenon of clay translocation concomitant with podzolization material was observed also in other Spodosols (Wang and McKeague 1982; Hseu et al. 2004).

The soil chemical properties of TS pedon are shown in Table 8.4. The pH value of the pedon increases with soil depth and ranges from 3.2 to 4.7 (H_2O) and ranges from 2.8 to 4.3 (1 N KCl), respectively. The maximum value of the soil organic carbon content is significantly found in the organic (Oe) layer and spodic horizon (Bhs) and decreases with soil depth below the spodic horizon. This indicates that organic C has eluviated from the surface organic horizon and accumulated in the spodic horizon. The exchangeable bases (K, Na, Ca, Mg) capacity are very low. The cation exchangeable capacity (CEC) is related primarily to the amount of soil organic matter, amorphous materials, and clay content distributed in the soil pedon. Because these materials are eluviated from upper horizon, the CEC values are higher in spodic horizon than in E horizon. Very low base saturation percentage (<5 %) also indicates that strong leaching processes are associated with the process of podzolization.

The climate is relatively cool and humid in this area. The vegetation type is mainly conifer forest. High precipitation caused a strong leaching process that was responsible for the low contents of exchangeable cations and low soil pH in the upper parts of Spodosol. Moreover, the decomposition of conifer forest organic residues, which are known to release high amount of organic acid, further decreased the soil pH.

The concentrations of the three forms of extractable Fe and Al determined by DCB (Fe_d and Al_d), ammonium oxalate (Fe_o and Al_o), and sodium pyrophosphate (Fe_p and Al_p) extraction methods are shown in Table 8.5. The concentrations of three different extractable Fe are normally higher in the B horizon compared to either E or A horizon. The contents of all three extractable aluminum forms are similar to iron forms, and they are significantly increased, as expected, in the spodic horizons compared to the E horizon. The concentrations of pyrophosphate extractable Fe and Al of the Bhs horizon are higher than those of E horizon. These results implied that the translocation of organo-Fe (or Al) complexes has occurred in the pedon. The general order of Fe is normally shown as: $Fe_d > Fe_p > Fe_o$, but there is no significant trend for Al content in these three extractions. This result is different from the order among the three extractable Fe and Al were $Fe_d > Fe_o > Fe_p$ and $Al_d > Al_o > Al_p$ in cool tropical Spodosols in Rwanda (Van Ranst et al. 1997). The ratio of Fe_p/Fe_o of spodic horizon is about 1.0. This ratio indicates that 50 % of the iron existed as organometallic complexes in the selected soils (Liu and Chen 1998, 2004). The ratio of Al_p/Al_o has a similar trend to that of Fe_p/Fe_o. This may indicate that the Fe and Al are mainly bound with organic matter. However, the ratio of Al_p/Al_o tends to be highly variable.

The distribution of organic C, Fe_p, Al_p, ODOE, and $Al_o + 1/2Fe_o$ indicates that the podzolization occurred in this area. Proposed chemical criteria for spodic material meet the oxalate-extractable materials ($Al_o + 1/2 Fe_o \geq 0.5$ % and ODOE value ≥ 0.25), and the value of the illuvial B horizon must be at least two times of the overlying eluvial E horizon (Rourke et al. 1988; Soil Survey Staff 2006). The above

Table 8.3 The physical properties of TS pedon in Chiayi County, central Taiwan

| Horizon | Depth (cm) | Total | | | Texture | Size class of sand[a] | | | | | Bd (Mg/m^3) |
		Sand (g/kg)	Silt (g/kg)	Clay (g/kg)		VC (g/kg)	C (g/kg)	M (g/kg)	F (g/kg)	VF (g/kg)	
Oe	0–5	–	–	–	–	–	–	–	–	–	–
A	5–15	587	282	131	S	1	1	85	372	126	–
AE	15–20	806	155	38	LS	0	1	128	485	191	1.51
E	20–30	769	188	42	LS	1	2	109	525	134	1.55
Bhs	30–43	702	183	115	LS	0	3	87	471	142	1.26
Bw	43–48	707	164	130	LS	0	1	273	406	27	0.91
C	48.70	807	133	60	LS	1	0	99	563	144	–

[a] VC (very coarse): 2.0–1.0 mm; C (coarse): 1.0–0.5 mm; M (medium): 0.5–0.25 mm; F (fine): 0.25–0.1 mm; VF (very fine): 0.1–0.05 mm
Copyright by Hseu et al. (2004)

Table 8.4 The chemical properties of TS pedon in Chiayi County, central Taiwan

| Horizon | Depth (cm) | pH | | Organic C (g/kg) | Exchangeable base | | | | Sum of Cations (cmol (+)/kg soil) | CEC[a] (cmol (+)/ kg soil) | BS[b] (%) |
		H$_2$O	KCl		Ca (cmol (+)/kg soil)	Mg (cmol (+)/kg soil)	K (cmol (+)/kg soil)	Na (cmol (+)/kg soil)			
Oe	0–5	3.2	2.8	203	0.07	0.28	0.73	0.07	1.14	29.6	4
A	5–15	3.7	3.1	113	0.02	0.06	0.13	0.01	0.22	20.5	1
AE	15–20	3.7	3.6	9	0.01	0.01	0.02	0.01	0.05	3.30	2
E	20–30	3.7	3.8	5	n.d.	0.01	0.02	n.d.	0.03	2.50	1
Bhs	30–43	4.1	4.0	20	n.d.	0.03	0.05	0.01	0.09	15.0	1
Bw	43–48	4.3	4.1	19	n.d.	0.02	0.05	0.01	0.09	14.1	1
C	48.70	4.7	4.3	9	n.d.	0.01	0.03	n.d.	0.04	7.70	1

[a] Cation exchange capacity
[b] Base saturation
Copyright by Hseu et al. (2004)

Table 8.5 The contents and ratios of Fe and Al in TS pedon in Chiayi County, central Taiwan

Horizon	Depth (cm)	Fe$_o$ (g/kg)	Fe$_p$ (g/kg)	Fe$_d$ (g/kg)	Al$_o$ (g/kg)	Al$_p$ (g/kg)	Al$_d$ (g/kg)	Al$_o$ + 1/2Fe$_o$ (%)	ODOE
Oe	0–5	10.0	19.3	18.3	2.6	3.6	2.8	0.76	1.79
A	5–15	9.0	8.3	16.0	2.2	1.6	2.1	0.67	1.67
AE	15–20	1.3	0.9	1.5	0.2	0.2	0.2	0.09	0.29
E	20–30	1.4	0.9	2.5	0.4	0.4	0.5	0.11	0.24
Bhs	30–43	27.3	27.6	43.4	7.0	7.7	6.4	2.07	2.28
Bw	43–48	21.5	22.2	34.2	8.2	7.8	7.8	1.98	2.10
C	48.70	9.6	8.0	17.3	5.0	4.0	5.0	0.98	0.87

Copyright by Hseu et al. (2004)

mentioned criteria for the Bhs horizon also meet the chemical criteria, which suggest that the spodic materials occurred in the pedon. It is clear that the spodic materials and clay particles have simultaneously accumulated in the Bhs horizon.

8.2.2 Ultic Haplorthods

The profile described below is situated in A-lisan, which is an important subalpine area for forest resource conservation in Taiwan. The subalpine forest soils, at 1,500 to 3,000 m elevation, are derived from marine argillaceous sediments and interstratified sandstone and shale of late Miocene to early Pliocene age. The overstory vegetation is dominated by red cypress (*Chamaecyparis formosensis*), and to a lesser extent by pine (*Pinus armandii*). The understory vegetation contains *Yushania niitakayanensis*, *Polygonum Chinese* Linn, and *Miscanthus floridulus*. Based on the climatic data of the Alishan Metrological Station for the past three decades (Lin et al. 2002), the annual rainfall is about 4,000 mm and most of it falls from May to September. The mean annual air temperature is 10.6 °C. The soil moisture regime is udic or perudic and the soil temperature regime is mesic in the area.

Field morphology of the A-lisan pedon is described as follows. If not otherwise indicated, soil color is given for moist soil.

Oe	0–10 cm	Very dark gray (7.5YR 3/1) many very fine and fine roots; clear smooth boundary
A	10–14 cm	Black (7.5YR 2.5/1) sandy loam; moderate very fine and fine granular structures; friable, slight sticky and slight plastic; many very fine and fine roots; abrupt smooth boundary
E	14–23 cm	Black (7.5YR 7/1) sandy loam; massive structure; firm, slight sticky and slight plastic; some very fine and fine roots; clear wavy boundary
EB	23–40 cm	Black (7.5YR 7/1) sandy loam; massive structure; firm, slight sticky and slight plastic; some very fine and fine roots; diffuse wavy boundary
Bhs	40–54 cm	Brown (7.5YR 4/3) sandy clay loam; moderate very fine and fine angular blocky structures; friable, slightly sticky and slight plastic; some very fine and fine roots; diffuse smooth boundary
Bt	54–80 cm	Strong brown (10YR 5/6) sandy clay loam; moderate very fine and fine angular blocky structures; firm, sticky and slight plastic; some very fine roots; diffuse smooth boundary
C	>80 cm	Strong brown (10YR 5/8) sandy clay loam; weak very fine angular blocky structures; friable, slightly sticky and slight plastic; some very fine roots

Surface layers (Oe & A) of more than 10 cm thickness were found in the pedon (Fig. 8.5). The soil colors in all subsurface horizons with different degrees of spodic material accumulation are almost same as 7.5YR 5/6 which is defined as one of morphological criteria for spodic horizon (Bhs) in *Soil Taxonomy* (Soil Survey Staff 2006). Various clay coatings and pellet-like aggregates were found in the Bhs and Bt horizons (Fig. 8.6), which partly meet the criteria of spodic horizon (Bhs) (Li et al. 1998; Liu and Chen 2004). Eluvial–illuvial phenomena were also clearly evidenced by changes in soil color and soil texture throughout the soil profile. The colors were gray or light gray with a chroma of 1 in all eluvial horizons, but the colors were strong brown or brownish yellow in the subsurface horizons. However, all albic horizons (E and EB) were sandy loam, but the spodic or argillic horizons were sandy clay loam to clay. The strong brown spodic horizon is underlain by similarly colored Bt horizon. It was difficult to identify clay coatings by the naked eye, but discontinuous few and common clay coatings were found in the Bhs and Bt horizons which can be identified in the field by a hand lens.

Fig. 8.5 Field morphology of the A-lisan pedon in Chiayi County, central Taiwan. Copyright by Prof. Hseu

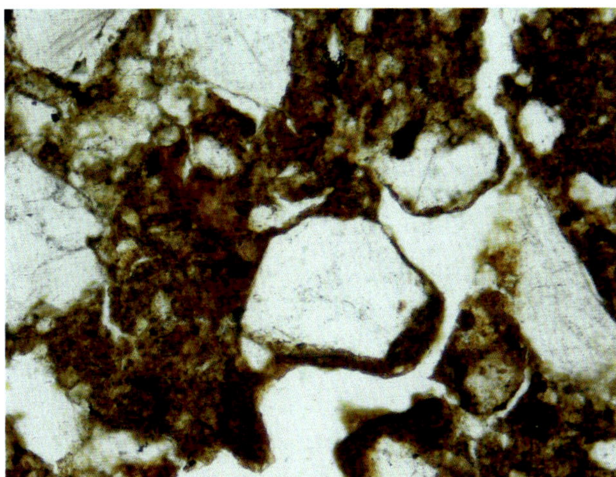

Fig. 8.6 Photomicrograph of organo-metal complex as grain coatings in the Bhs of the A-lisan pedon in Chiayi County, central Taiwan (PPL). The width of photo frame is 1.0 mm. Copyright by Prof. Hseu

Regarding micromorphology, the organic layer consists of flora fragments with varying degrees of decomposition. Fresh roots, humified substances mixed with broken cell walls, fecal pellets, or amorphous organic materials are found in the O and A horizons. While clean coarse grains dominate the micromorphological characteristics of the E horizon, microstructures are associated with trace amounts of packing voids and cracks, roots, and black humidified materials. The dominant pedofeatures of Bhs horizon were organometallic complexes as pellets or aggregate-like on the coarse grain and ped surface. Few moderately and well-oriented clay coatings were found on ped surfaces associated with dark pellet-like aggregates in the Bhs horizon. Surprisingly, organometallic and clay accumulations were identified both as pellets and clay coatings (Fig. 8.6). These

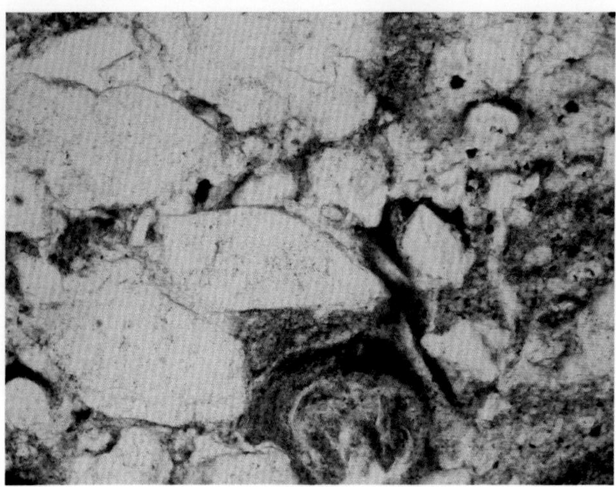

Fig. 8.7 Photomicrograph of clay coating residues in the Bt horizon of A-lisan pedon in Chaiyi County, central Taiwan (PPL). The width of photo frame is 0.625 mm. Copyright by Prof. Hseu

illuvial clays, however, have a relatively weak birefringence masked by amorphous materials. Similarly, both organo-metallic complexes and clay coatings were found in the Bt horizon.

Illuvial clays were seen as various forms in subsurface horizons, e.g., grain coatings, hypocoatings along the root channels, and infillings in larger packing voids (Fig. 8.7). These illuvial clays have a relatively weak birefringence caused by organometallic complexes, especially when in Bhs horizon. The skeleton grains were either uncoated or coated to different degrees with materials similar in appearance to the aggregates in the typical loamy Spodosols (Phillips and FitzPatrick 1999). These aggregates and coatings consisted of a mixture of fine mineral materials. Thick clay coatings in >1 % of the area in thin section, darkened by organometallic complexes, are found in all Bhs horizon. Based on the micromorphology, we can infer that the

translocation of clay occurred prior to the illuviation of organometallic complexes in the subsurface horizons in this area.

The pedon has fine textures in the subsurface horizons. The pH values of soils are less than 5.0 (Table 8.6), and those of Bhs and Bt horizons are significantly lower than the maximum pH value defined for spodic materials (pH $\leqq$ 5.9). Organic C contents are greatest in the surface organic horizon, decrease in the E horizon, and increase in the Bhs horizon. The accumulation of clay and organic carbon in the subsurface horizon was mainly caused by strong leaching potential from very high rainfall in the area. Cation exchange capacity followed the same trend as OC. Extremely low base saturation of the soils ranged from 7 to 10 % in the pedon because of dominantly acidic sandstone parent materials and strong leaching environment.

The concentrations of extractable Fe and Al are normally higher in the B horizons compared to either E or A horizon (Table 8.7). The orders of Fe and Al in this pedon are variable. The DCB-extractable Fe contents are much higher than oxalate and pyrophosphate extractable ones identified by the Fe_p/Fe_d and Fe_o/Fe_d values. Most pyrophosphate extractable Fe and Al contents were higher than oxalate ones within the same horizons. High ratios of Fe_p/Fe_o and Al_p/Al_o in the B horizons, more than 1.0, suggested that most of noncrystalline Fe and Al exists in organic-bound form. Bockheim et al. (1996) also showed the similar trend between these two extractions for Ultic Spodosols in Oregon, but their average ODOE values of the subsurface horizons were less than those of A-lisan pedon.

Combinations or ratios of extractable Fe and Al can be used to interpret the degree of podzolization. Accumulation of spodic materials was identified by the values of $Al_o + 1/2 Fe_o$ (%) and ODOE (De Coninck 1980). In A-lisan pedon, these two parameters satisfied the criteria defined in spodic material based on *Soil Taxonomy* (Soil Survey Staff 2006), and indicated that the strong podzolization processes occurred. The

Table 8.6 The physical and chemical properties of the A-lisan pedon in Chaiyi County, central Taiwan

Horizon	Bd[a] (Mg/m^3)	Texture			pH	OC[b] (g/kg)	CEC[c] (cmol(+)/kg)	BS[d] (%)
		Sand (g/kg)	Silt (g/kg)	Clay (g/kg)				
Oe	−[e]	−	−	−	4.4	82	19.7	10
A	−	604	254	142	3.9	31	14.3	10
E	−	635	240	126	4.4	8	14.1	9
EB	1.6	574	254	171	4.5	10	17.0	7
Bhs	1.1	490	232	288	4.5	26	21.6	7
Bt	−	454	266	280	4.5	22	19.1	7

[a] Bulk density
[b] Organic carbon
[c] Cation exchangeable capacity
[d] Base saturation percentage
[e] No determined
Copyright by Hseu et al. (2004)

Table 8.7 The contents and ratios of Fe and Al of the A-lisan pedon in Chaiyi County, central Taiwan

Horizon	Fe_d^+	Al_d (g/kg)	Fe_o^+ (g/kg)	Al_o (g/kg)	Fe_p^+ (g/kg)	Al_p (g/kg)	$Al_o + (1/2) Fe_o$ (%)	Fe_p/Fe_o (%)	Fe_p/Fe_d	Fe_o/Fe_d	Al_p/Al_o	ODOE[+]
Oe	–	–	–	–	–	–	–	–	–	–	–	0.66
A	18.8	0.9	2.80	0.97	1.34	0.86	0.24	0.48	0.07	0.15	0.88	0.62
E	18.4	0.4	0.68	0.38	0.28	0.28	0.07	0.41	0.02	0.04	0.73	0.20
EB	33.3	0.9	1.78	1.08	1.90	0.83	0.20	1.07	0.06	0.05	0.76	0.11
Bhs	61.2	12.9	14.5	4.96	17.3	5.16	1.22	1.20	0.28	0.24	1.04	2.35
Bt	72.5	7.2	11.6	5.34	19.3	5.58	1.12	1.66	0.27	0.16	1.05	1.13

[+] Subscripted d, o and p are citrate-dithionite, oxalate and pyrophosphate extractable
[+] Optical density of oxalate extraction
[–] No determined
Copyright by Hseu et al. (2004)

value of Al_o + 1/2 Fe_o (%) in the Bhs horizon is much higher than twice of those in the upper layers. The depth functions of ODOE value correlated well with OC content in this pedon. The spodic horizon has reddish hues (7.5YR) and high levels of oxalate-extractable Fe and Al and illuvial OC as reflected by the ODOE values in excess of 0.25. The ODOE test was designed to indicate the presence of fulvic acids that are crucial in the translocation of Fe and Al in Spodosols (Daly 1982). Therefore, fulvic acids might play a role as major chelating agents for transporting Fe and Al associated with relatively high ODOE values in the Bhs horizons (Hseu et al. 2004).

The Bt and Bhs horizons satisfied the chemical criteria defined in spodic horizon from the keys to *Soil Taxonomy*. Hseu et al. (2004) found that pyrophosphate extractable Fe and Al values correlated well with OC values ($r^2 = 0.71$ and 0.38, $p < 0.05$) for all the eluvial and illuvial horizons (n = 15) of three pedons in this area, further indicating that the clay content correlated well not only with pyrophosphate extractable Fe and Al ($r^2 = 0.68$ and 0.74, $p < 0.05$), but also with dithionite citrate extractable Fe and Al ($r^2 = 0.65$ and 0.73, $p < 0.05$). The high r^2 values suggest that illuvial Fe and Al are significantly bound with organic matter and then leached downward with clay. Therefore, although these organometallic complexes may adsorb or precipitate within the finer matrix formed by the illuvial particles, they only move down for a short distance.

Regarding clay mineralogy, vermiculite, vermiculite-illite interstratified minerals, illite, hydroxy-interlayer vermiculite (HIV), kaolinite, quartz and gibbsite with variable quantities are present in the Spodosols of A-Lishan area (Lin et al. 2002). Vermiculite-illite interstratified minerals were characterized by the basal XRD peak intermediated between 1.0- and 1.4-nm for Mg-saturated clays (Fig. 8.8). Glycerol treatment does not alter the spacings, but on K-saturated and heating to 100 °C the layers of the vermiculite component collapse to the illite spacing, resulting in a 1.0 nm peak and a series of higher orders. The broad peaks in this range indicated vermiculite-illite interstratified minerals in the O/A horizons. The XRD patterns of Mg-saturated clay at 25 °C

indicated that illite (1.0- and 0.334-nm peaks), kaolinite (0.72-nm peak), and quartz (0.426- and 0.334-nm peaks) were present. Gibbsite was reconfirmed by the peak (0.485-nm) which disappeared when the K-saturated clays were heated to 350 °C. The HCl treatment without cationic saturation was used to destroy the chlorite structure and to differentiate chlorite and kaolinite which have very different structures and pedogenic occurrences, but the difficulties exist when they present in mutual mixtures. Weak reflections at 25 °C are present at 12.5 and 25° (2θ), and they could contain only chlorite, only kaolinite, or a mixture of the two. However, the residual peaks at 12.5 and 25° (2θ) indicated the only presence of kaolinite, but not chlorite.

Significantly broad peaks between 1.0- and 1.4-nm of the XRD patterns of Mg-saturated clays indicated that vermiculite-illite interstratified minerals were present in the Bhs horizon. These broad peaks between 1.0- and 1.4-nm by the K-saturation biased to 1.4-nm at room temperature, and completely collapsed to 1.0-nm at 350 °C. Hence, more vermiculite was identified than HIV. It is noted that no smectite was identified because the 1.4-nm peak of the Mg-saturated sample does not expand to 1.7-nm treated with glycerol, because that the extremely low base saturation and good drainage conditions resulted in no smectite formation. Minor amounts of kaolinite and gibbsite are present through the soil profiles. In consequence, Lin et al. (2002) proposed that the weathering sequence of clay minerals in these soils is: illite $\rightarrow$ vermiculite (or vermiculite-illite interstratified minerals) $\rightarrow$ HIV and vermiculite.

8.3 Formation and Genesis

Several processes have been proposed to explain podzolization: (i) formation and downward transport of complexes of organic acids with Al and Fe (De Coninck 1980), (ii) silicate weathering followed by downward transport of Al and Si as inorganic colloidal sols (Jakobsen 1991), and (iii) Al, and probably Fe, is first translocated into the B horizon as proto-imogolite sols (Farmer 1982). While theories of

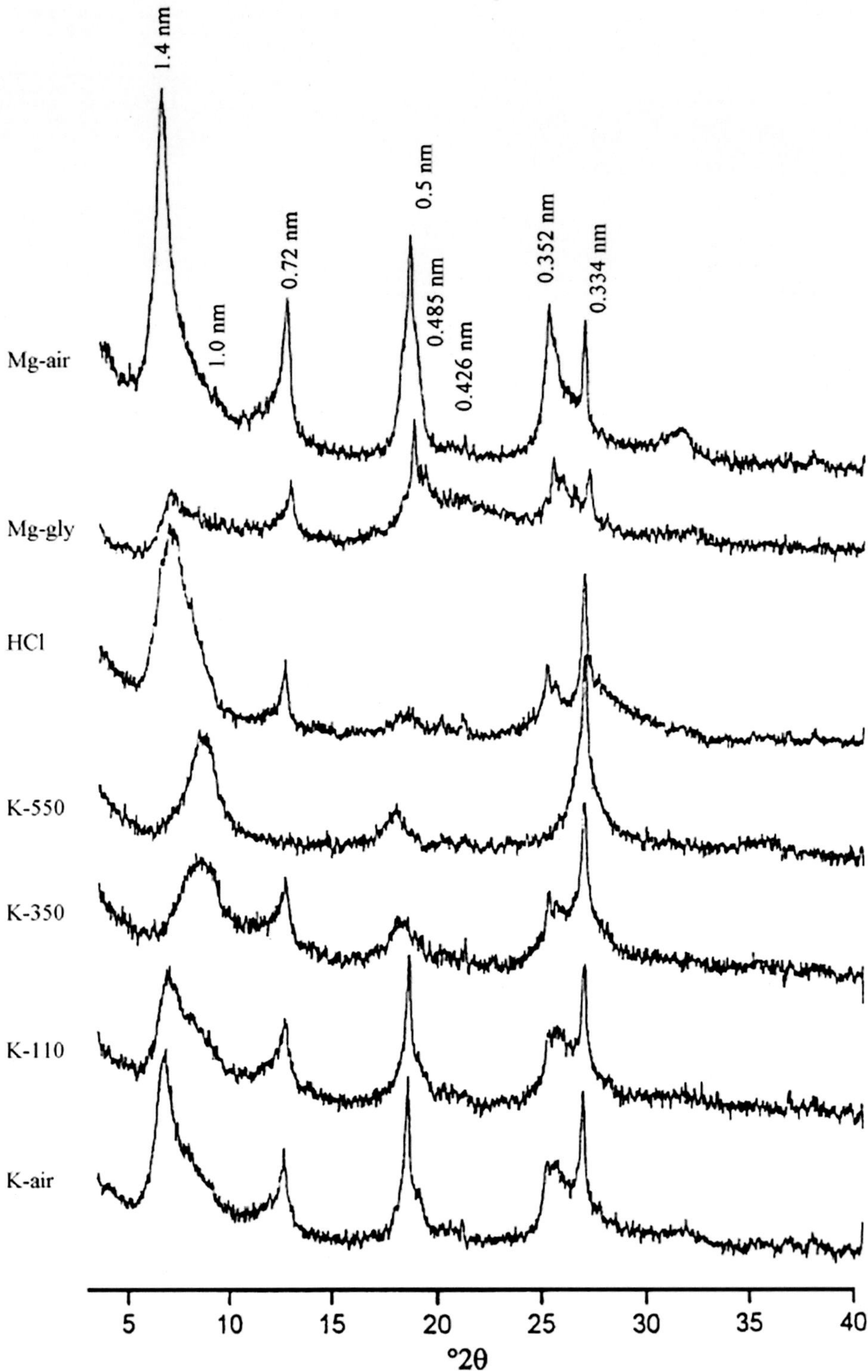

Fig. 8.8 The XRD patterns of clay fractions in the Bhs horizon of the A-lisan pedon in Chaiyi County, central Taiwan. Copyright by Prof. Hseu

illuviation of spodic materials have largely emphasized chemical processes rather than clay translocation, some studies demonstrated clay particle as well as chemical migration of organo-metal complexes in podzolic soils (Guillet et al. 1975; Alekseyev 1983; Miller 1983; Harris and Hollien 1999).

Two commonly used pedogenic pathways resulted in distinctly different soils as Ultisols with spodic (Bhs) horizons or Spodosols with argillic (Bt) horizons (Markewich and Pavich 1991). According to Bockheim et al. (1992), well-drained soils progressively developed from Typic Troporthods to Typic Haplohumults on the marine terraces in southwestern Oregon. Guillet et al. (1975) indicated that illuviation of clay can promote differentiation of podzol horizon and podzolization is strongly redistributive in a Podzol which demonstrated the clay accumulation in the upper parts of spodic horizon under a texture class of sand and a heath vegetation.

Significant accumulation of organic matter, Fe, Al, and clay in the Bhs horizons of Ultic Haplorthods in Taiwan indicates that podzolization processes and clay illuviation were occurred simultaneously in the soils. Mobile organic carbon from plant residue was leached down and complexed with Fe and/or Al in the albic horizons (Hseu et al. 2004). These organometallic complexes were immobilized at some depth through supplementary fixation of cations as suggested by De Coninck (1980). The coarser soil texture facilitated the leaching of organic matter from surface horizons into B horizons and subsequent deposited at lower parts of the pedon.

The significant higher clay content in the Bhs horizons than E horizons may indicate that the clay has significantly eluviated from E horizon and been deposited in the Bhs horizons or mixed with organometallic complexes in the Bhs horizons. Clay illuviation is commonly recognized by the occurrence of argillans (Soil Survey Staff 2006). In these soils, however, only poor-oriented argillans were occurred in Bhs horizons. Wang and McKeague (1982) pointed out that translocation of clay complexed with organic matter, Fe, and Al might cause a lack of well-oriented clay in the spodic (Bhs) horizons despite occurrence of large amounts of illuviated clay in the Bhs horizons. The pedogenic processes in the study area are characterized mainly by organometallic complexes and illuvial clay in the spodic horizons. The large amounts of spodic materials and clay in Bhs horizons indicate that podzolization and clay illuviation (eluviated from E horizons) were two dominant pedogenic processes that occurred in these Spodosols. Higher contents of clay and Fe_d in the Bhs horizons seem to be the main reason that soil colors are yellower than those of reddish color of typical Spodosols in the world.

8.4 Land Use and Management

Typical Spodosols are generally found in the subalpine forests of Central Range, Taiwan, where coniferous trees are the major plants in the canopy level. These coniferous trees had been a main timber source in Taiwan before the 1980s. To conserve natural resource and sustain forest ecosystem after this times, the Spodosols were applied to afforestation. Therefore, the soil properties are crucial for seedling growth like soil acidity, base saturation, drainage, and clay content.

Information about the soil survey conducted on high mountain areas is helpful for afforestation, which is introduced in Chap. 2.

For ecological management of forest resource, Spodosols have been considered a record of forest evolution by the formation processes of spodic materials. For example, the mean SOC stock of Spodosols is higher (41.9 kg/m^2) than other soil orders in the upper 100 cm depth. The higher mean values of soil organic carbon of Spodosols is attributed to the large number of complexes of soil organic matter in the soil surface (Chen and Hseu 1997) or spodic horizon containing higher illuvial humus and iron materials, and thicker organic layer in the soil surface (Vejre et al. 2003). However, the mean soil carbon stock of forest soils in Taiwan, excluding Histosols and Spodosols, is 10.1 kg/m^2 (0–30 cm depth), 13.8 kg/m^2 (0–50 cm depth), and 18.5 kg/m^2 (0–100 cm depth). Over-removal of forest trees causes accelerated loss of SOC from Spodosols, so that the Taiwan government currently encourages afforestation in these areas for sustainable natural resource management.

References

Alekseyev VY (1983) Mineralogical analysis for the determination of podzolization, lessivage and argillation. Sov Soil Sci 15:21–28

Bockheim JG, Kelsey HM, Marshall JG III (1992) Soil development, relative dating, and correlation of late quaternary marine terraces in southwestern Oregon. Quatern Res 37:60–74

Bockheim JG, Marshall JG III, Kelsey HM (1996) Soil-forming processes and rates on uplifted marine terrace in southwestern Oregon, USA. Geoderma 73:39–62

Chen ZS, Hseu ZY (1997) Total organic carbon pool in soils of Taiwan. Proc Nat Sci Counc ROC. Part B: Life Sci 21:120–127

Chen ZS, Liu JC, Chiang HC (1995) Soil properties, clay mineralogy, and genesis of some alpine forest soils in Ho-Huan Mountain area of Taiwan. J Chin Agric Chem Soc 33:1–17

Chen ZS, Lin KC, Chang JM (1989) Soil characteristics, pedogenesis, and classification of Beichateinshan Podzolic soils, Taiwan. J Chin Agric Chem Soc 27:145–155 (in Chinese, with English abstract)

Daly BK (1982) Identification of Podzol and podzolized soils in New Zealand by relative absorbance of oxalate extracts of A and B horizons. Geoderma 28:29–38

De Coninck F (1980) Major mechanisms in formation of spodic horizon. Geoderma 24:101–128

Farmer VC (1982) Significance of the presence of allophane and imogolite in Podzol Bs horizons for podzolization mechanism: a review. Soil Sci. Plant Nutr. 28:571–578

Guillet BJ, Rouiller T, Souchier B (1975) Podzolization and clay migration in Spodosols of eastern France. Geoderma 14:223–245

Harris WG, Hollien KA (1999) Changes in quantity and composition of crystalline clay across E-Bh boundaries of Alaquods. Soil Sci 164:602–608

Hseu ZY, Tsai CC, Lin CW, Chen ZS (2004) Transitional soil characteristics of Ultisols and Spodosols in the subalpine forest of Taiwan. Soil Sci 169:457–467

Hseu ZY, Chen ZS, Wu ZD (1999) Characterization of Placic Horizons in two subalpine forest inceptisols. Soil Sci Soc Am J 63:941–947

Jakobsen BH (1991) Multiple processes in the formation of subarctic podzols in Greenland. Soil Sci 152:414–426

Li SY, Chen ZS, Liu JC (1998) Subalpine loamy Spodosols in Taiwan: characteristics, micromorphology, and genesis. Soil Sci Soc Am J 62:710–716

Lin CW, Hseu ZY, Chen ZS (2002) Clay mineralogy of Spodosols with high clay contents in the subalpine forests of Taiwan. Clays Clay Miner 50:726–735

Liu JC, Chen ZS (1998) Soil characteristics and pedogenesis of loamy Spodosols near Wang-Hsiang mountains in central Taiwan. J Chinese Agric Chem Soc 36:151–162

Liu JC, Chen ZS (2004) Soil Characteristics and clay mineralogy of two subalpine forests Spodosols with clay accumulation in Taiwan. Soil Sci 169:66–80

Lyford WH (1952) Characteristics of some podzolic soils of northern United States. Soil Sci Soc Am Proc 16:231–234

Markewich HW, Pavich MJ (1991) Soil chronosequence studies in temperate to subtropical, low-latitude, low-relief terrain with data from the eastern United States. Geoderma 51:213–239

McKeague JA, De Connick F, Franzmeier DP (1983) Spodosols. In: Wilding LP, Smeck NE, Hall GF (eds) Pedogenesis and Soil Taxonomy, II. Elsevier, New York, pp 217–252

Miller BJ (1983) Ultisols. In: Wilding LP, Smeck NE, Hall GF (eds) Pedogenesis and Soil Taxonomy: II. The Soil Orders. Elsevier, New York, pp 283–323

Phillips DH, FitzPatrick EA (1999) Biological influences on the morphology and micromorphology of selected Podzols (Spodosols) and Cambisols (Inceptisols) from the eastern United States and north-east Scotland. Geoderma 90:327–364

Rourke RV, Brasher BR, Yeck RD, Miller FT (1988) Characteristics and morphology of U. S. Spodosols. Soil Sci Soc Am J 52:445–449

Soil Survey Staff (2006) Keys to soil taxonomy, 10th edn. USDA-SCS, Washington, DC

Sommer M, Halm D, Geisinger C, Andruschkewitsch I, Zarei M, Stahr K (2001) Lateral podzolization in a sandstone catchment. Geoderma 103:231–247

Vejre H, Calleses I, Vesterdal L, Raulund-Rasmussen K (2003) Carbon and nitrogen in Danish forest soils-Contents and distribution determined by soil order. Soil Sci Soc Am J 67:335–343

Van Ranst E, Stoops G, Gallez A, Vandenberghe RE (1997) Properties, some criteria of classification and genesis of upland forest Podzols in Rwanda. Geoderma 76:263–283

Wang C, McKeague JA (1982) Illuvial clay in sandy podolic soils of New Brunswick. Can J Soil Sci 62:79–89

9.1 General Information (Soil Formation Factors and Distribution of Great Group)

Oxisols and Ultisols generally form in warm and humid conditions in tropical or subtropical regions. Their subsurface horizons are commonly red or yellow in color, evidencing the accumulations of iron oxides. Landscapes associated with some of these soils are stable and old, measured in hundred thousands of years, and the thickness can be up to 5 m in Taiwan (Fig. 9.1). The age obtained from an Oxisol pedon on the Quaternary terrace of aged alluvium in Central Taiwan, approximate 400,000 years estimated by beryllium-10, may represent the minimum time required for soils develop into Oxisols in Taiwan (Tsai et al. 2008). Based on the the crystallinity ratios of pedogenic iron, the soils give an estimated minimum in age of forty thousand years for Ultisols on the Quaternary terraces in Taiwan (Tsai et al. 2006, 2007). Plinthite may be present resulted from alternative wet and dry cycles by fluctuating water tables (Fig. 9.2). However, plinthite only occurs in the Ultisols with seasonal high water tables on the Quaternary terraces of aged alluvium in Northern and Central Taiwan, particularly on lowland rice (paddy rice) fields (Hseu and Chen 1997, 2001; Jien et al. 2010). Oxisols of Taiwan are generally well-drained and have no plinthite.

Due to the moist conditions in warm to tropical climates of Taiwan, Ultisols are extensively found on the stable and old land surfaces such as flat hill forests and Quaternary terraces of aged alluvium, while Oxisols are locally on the highest level of the Quaternary terraces (Fig. 9.3). Ultisols are dominantly characterized by clay mineral weathering, migration of clays to accumulate in the subsurface horizon, and leaching of base cations from the profile. Ultisols are typical soils with strong acidity and less than 35 % base saturation at depth. OXISOLS are the most highly weathered soils in Soil Taxonomy and the WRB system. Ultisols have either an argillic horizon or kandic horizon, while Oxisols have an either oxic horizon or a surface horizon with 40 % clay that overlies a highly weathered kandic horizon with few weatherable minerals remaining.

Ultisols are about 9.6 % (1,624 km^2) of rural soils in Taiwan, while Oxisols are only <1 % (50 km^2) of total rural soils. In forests, Ultisols are about 7.50 % (1,250 km^2) of the total forest soils, and no Oxisol can be found herein (Guo et al. 2009). Hapludoxs and Kandiudoxs are only two recognized Great Groups of Oxisols in Taiwan, but some of the highly weathered Ultisols are marginal to the criteria of Oxisols. Only three soil series of agricultural land have been confirmed as Oxisols in Taiwan, which are always for tea or pineapple production in land use. Seven Great Groups of Ultisols have been recognized in Taiwan. They are Endoaquults, Paleaquults, Plinthaquults, Haplohumults, Hapudults, Paleudults, and plinthudults. Moreover, Endoaquults, Paleaquults, Plinthaquults, and plinthudults are commonly found on the Quaternary terrace of aged alluvium in Northern and Central Taiwan, which varied with soil-moisture regime controlled by land use particularly in paddy rice production. For example, the long-term fluctuation of seasonal high water table in these areas can cause not only the formation of plinthite but also poor drainage in the deep Ultisols. Compared to the deep Ultisols, Haplohumults and Hapudults are only distributed stable hilly areas through Taiwan such as lowland rain forests and subapline forests regardless with parent materials (Tsai and Chen 2000; Tsui et al. 2004; Guo et al. 2009; Cheng et al. 2011).

9.2 Profile Characteristics (Morphology and Micromorphology, Physiochemical Properties, and Mineralogy)

Two pedons of Hapludoxs, pedons YM and PK1, are introduced in this chapter. Pedon YM comes from the second highest levels of the Taoyuan terraces which are the series of river terraces with six levels and have been developed by clockwise migration (from westward to northeastward) of the Tahan River in Northern Taiwan

Z.-S. Chen et al., *The Soils of Taiwan*, World Soils Book Series,
DOI 10.1007/978-94-017-9726-9_9, © Springer Science+Business Media Dordrecht 2015

Fig. 9.1 Typical ultisol profile in Taiwan. Copyright by Prof. Hseu

Fig. 9.2 Ultisols with plinthite on the paddy fields of Northern Taiwan. Copyright by Prof. Hseu

(Chen and Liu 1991). The second Oxisol, Pedon PK1, is located on the highest geomorphic surface of the Pakua tableland in Central Taiwan which consist of six levels of wide and unpaired surfaces developed by the Choushui River and its tributaries (Yang 1986). This tableland consisting of Quaternary terrestrial sediments was asymmetrically folded as an active anticline (above a décollement in Miocene rocks at a depth of more than 4 km) by frontal thrusting of the Changhua Fault system in response to crustal shortening. The southern part of the Pakua tableland consists of a series of very wide unpaired terraces. The five levels of terraces in order from PK1 (highest, 440 m above the sea level) to PK5 (lowest, 300 m above the sea level), are introduced herein because their soils are all Oxisols and Ultisols. However, the soils on the lowest level (PK6) are Inceptisols which are excluded in this chapter.

9.2.1 Morphology of Oxisol

9.2.1.1 Pedon YM

Field morphology of Pedon YM is described as follows. If not otherwise indicated, soil color is given for moist soil. The Pedon YM comprises thick and well developed oxic

(Bo) horizons of reddish hue, and moderate to strong subangular or angular blocky structures (Fig. 9.4).

Ap1	0–10 cm	Yellowish red (5YR 4/6) silty clay loam; moderate course angular blocky structure; firm, slight sticky and slight plastic; some fine roots; diffuse smooth boundary
Ap2	10–22 cm	Yellowish red (5YR 4/6) silty clay loam; moderate course angular blocky structure; firm, slight sticky and slight plastic; few very fine roots; diffuse smooth boundary
Bo1	22–42 cm	Reddish brown (5YR 4/4) clay; moderate medium subangular blocky structure; firm, sticky, and plastic; very few and very fine roots; diffuse smooth boundary
Bo2	42–73 cm	Reddish brown (2.5YR 4/4) clay; moderate fine and medium subangular blocky structures; very firm, sticky and plastic; very few and very fine roots; diffuse smooth boundary
Bo3	73–108 cm	Reddish brown (2.5YR 4/4) clay; moderate fine and medium subangular blocky structures; very firm, sticky and plastic; very few and very fine roots; diffuse smooth boundary

(continued)

(continued)

| Bo4 | 108–128 cm | Red (2.5YR 4/6) clay; moderate fine and medium subangular blocky structures; very firm, sticky and plastic; diffuse smooth boundary |
| Bo5 | 128–146 cm | Red (2.5YR 4/6) clay; moderate fine and medium subangular blocky structures; very firm, sticky and plastic; diffuse smooth boundary |

9.2.1.2 Pedon PK1

Field morphology of Pedon PK1 pedon is described as follows. If not otherwise indicated, soil color is given for moist soil (Fig. 9.5). The oxic (Bo) horizons have hues of 2.5YR which are much redder than those in the other subsurface horizons on the Pakua Tableland. Fine textures of the soils are reflected in their structure and consistence. In the Bo horizons, dominant groundmass can be found as isotropically oxidic materials, with less extent of weakly stipple clay coatings (Fig. 9.6). Development of structure and degree of stickiness and plasticity, as well as the abundance of clay coating,

Fig. 9.4 Profile morphology of YM. Copyright by Prof. Hseu

increase with soil age on the Pakua Tableland. Strongly coarse angular and subangular blocky structures are generally found in the soil profile. High clay content of the soils is also associated with the very sticky and plastic consistence.

A	0–30 cm	Yellowish red (5YR 5/8) silty clay; strong fine, medium, and coarse subangular blocky structures; firm, very sticky, and plastic; many very fine and fine roots; some coarse and very coarse biopores; few very fine and fine voids; smooth and diffuse boundary
Bo1	30–70 cm	Red (2.5YR 4/8) silty clay; strong medium, coarse and very coarse subangular blocky structures; firm, very sticky and plastic; many very fine and fine roots; few coarse and very coarse biopores; few very fine and fine voids; smooth and diffuse boundary
Bo2	70–110 cm	Red (2.5YR 4/8) silty clay; strong medium, coarse and very coarse subangular blocky structures; firm and very firm, very sticky and plastic; few very fine roots; few very fine biopores; few very fine voids; smooth diffuse boundary

(continued)

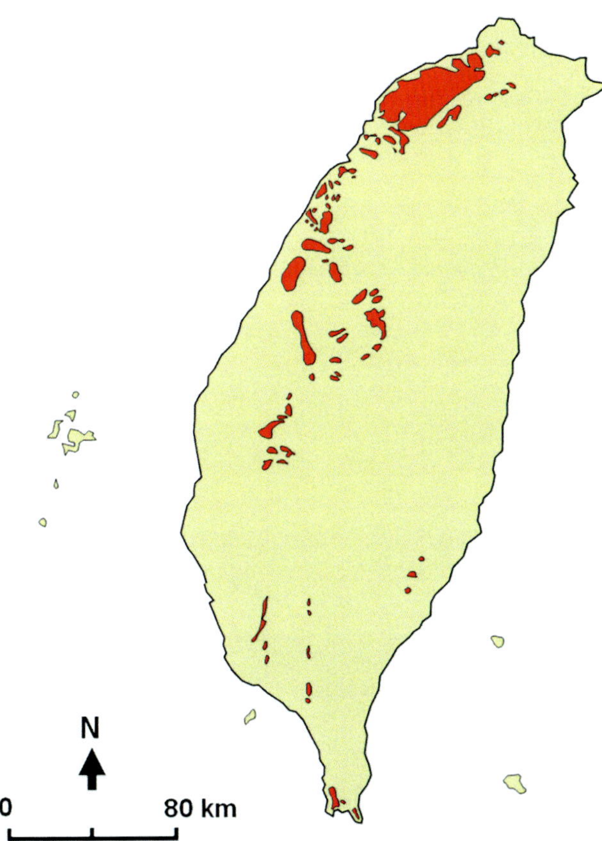

Fig. 9.3 Distribution of main Ultisols and Oxisols in Taiwan. Copyright by Prof. Hseu

(continued)

Bo3	110–140 cm	Red (2.5YR 4/6) silty clay; strong very coarse subangular blocky structures; firm and very firm, very sticky and plastic; few very fine roots; few very fine biopores; few very fine voids; smooth diffuse boundary
Bo4	140–160 cm	Red (2.5YR 4/6) silty clay; medium coarse and very coarse subangular blocky structures; firm, very sticky and plastic; few very fine roots; few very fine biopores; few very fine voids; smooth diffuse boundary
Bo5	160–200 cm	Red (2.5YR 4/6) silty clay; strong medium coarse and very coarse subangular blocky structures; firm and very firm, very sticky and plastic; few very fine roots; few very fine biopores; few very fine voids; smooth diffuse boundary

9.2.2 Physical and Chemical Properties and Clay Mineralogy of Oxisol

Relatively low abundances of sand and high abundances of silt and clay are found in the particle-size distribution for pedons YM and PK1 (Table 9.1). All sand fractions are fine to very fine. All the pH values are less than 5.0, which are corresponding to low organic carbon contents and poor CEC in the soils. The diagnostic horizons of pedons YM and PK1 meet the requirement for low activity clays (<16 cmol/kg) as defined in the USDA *Soil Taxonomy* (Soil Survey Staff 2010). Clear amounts of gibbsite and kaolinite are found throughout all the horizons in the pedons, and they are associated with high Fe_d/Fe_t ratios. In pedons YM and PK1, Kaolinite, illite, and quartz are the predominant minerals while chlorite and mectite are absent. Additionally, vermiculite, vermiculite-illite interstratified minerals, and gibbsite are also common in the Oxisols in variable quantities.

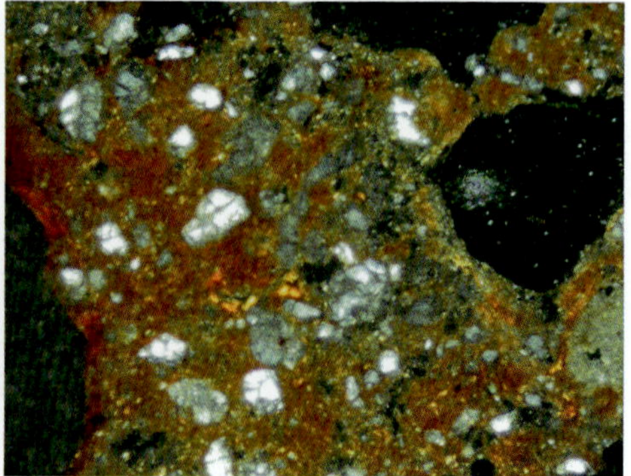

Fig. 9.5 Field morphology of pedon PK1 on the on the Pakua tableland in Central Taiwan. Copyright by Prof. Hseu

9.2.3 Morphology of Ultisol

9.2.3.1 Pedon PK2 on Quaternary Terrace of Aged Alluvium

Soil color was shown in moist (Fig. 9.7), except special explanation. With respect to micromorphology, translocated clays are clearly shown as reddish coatings on the pedsurface of Bt horizons (Fig. 9.8).

Fig. 9.6 Microphotograph of Bo horizons from Pedon PK1 by plane polarized light (*left*) and cross polarized light: oxic groundmass with weakly stipple clay coatings in the Bo1 horizon of PK-1 pedon, the width of frame is 1.5 mm. Copyright by Prof. Hseu

A	0–20 cm	Strong brown (7.5YR 4/6) silty clay loam; strong fine and medium subangular blocky structures; sticky and plastic; many medium and coarse roots; some fine and medium biopores; few very fine voids; diffuse wavy boundary
Bt1	20–51 cm	Strong brown (7.5YR 5/8) silty clay; strong medium and coarse subangular blocky structures; few and distinct clay coatings on the pedsurfaces; sticky and plastic; many medium and fine roots; many medium and coarse biopores; few very fine voids; diffuse wavy boundary
Bt2	51–85 cm	Yellowish red (5YR 5/6) silty clay; strong fine and medium angular blocky structures; common and distinct clay coatings on the pedsurfaces; very sticky and plastic; very fine roots; some coarse and medium biopores; few very fine voids; smooth diffuse boundary
Bt3	85–120 cm	Yellowish red (5YR 5/6) silty clay; strong medium and coarse angular blocky structures; common and distinct clay coatings on the pedsurfaces; very sticky and plastic; some fine roots; some fine and very fine biopores; few very fine voids; smooth diffuse boundary
Bt4	120–145 cm	Yellowish red (5YR 5/6) silty clay; strong medium and coarse angular blocky structures; common and distinct clay coatings on the pedsurfaces; very sticky and plastic; few very fine and fine biopores; few very fine and fine voids; smooth diffuse boundary
Bt5	>145 cm	Yellowish red (5YR 5/8) silty clay; strong medium and coarse subangular blocky structures; few and faint clay coatings on the pedsurfaces; very sticky and plastic, >1 % clay skins; few very fine biopores; few very fine voids

9.2.3.2 Pedon Liu2 on Coral Reef Terrace

"Terra Rossa" is specifically designated for the red (enriched in iron-oxides) and clayey soils developed from limestone or coral reef in the warm and humid climate along the Mediterranean Sea (Macleod 1980; Yaalon 1997; Liao et al. 2006).

Liuchiu Island is a raised coral-reef island about 14 km off southwestern coast of Taiwan. A thin bed of gently tilting reef limestone of Middle-Late Pleistocene age outcrops most of the island. Different surfaces on the marine terraces were produced from Late Pleistocene to Holocene. The surfaces of the upper two levels from Late Pleistocene terraces are covered by Ultisols. Instead of soil development entirely from local alluvium deposits, a recent argument suggested the Ultisols in Liuchiu Island had received eolian dusts from the continental China as part of soil materials (Cheng et al. 2011). The Pleistocene terraces were found in Liuchiu Island, and were tilted with the altitude at their peaks range 67–87 m

above the sea level. It serves as evidence that the island was uplifted and tilted simultaneously throughout the years. Based on ^{14}C dating (Shih et al. 1991), the island is uplifted at the rate 0.15–0.45 mm/year, and the island is continuing uplift with due to tectonic compression from the east compressive setting. The ages of the Pleistocene terrace estimated by ^{14}C dating of nearby coral limestone ranged from 31,600 to 34,000 year BP (Shih et al. 1991). All these age estimated are maximum soil ages; soil development may have begun some time after uplift of the marine terraces. The field morphological characteristics of Pedon Lu2 on the coral reef terrace of Liuchiu Island are described as follows (Fig. 9.9):

Ae	0–10 cm	Dark yellowish brown (10YR 3/4) silty clay; moderate fine and medium gradular parting to subangular blocky structures; moderate firm, very sticky and very plastic; many fine roots; many fine and medium biopores; clear smooth boundary
AB	10-30 cm	Dark yellowish brown (10YR 4/6) silty clay; moderate fine and medium subangular blocky structures; thick clay infillings along root channels; moderate firm, very sticky and very plastic; some fine roots and biopores; wavy diffuse boundary
Bt1	31–50 cm	Yellowish brown (10YR 5/6) clay; moderate fine and medium subangular blocky structures; few medium clay coatings; moderate firm, very sticky and very plastic; some fine roots and biopores; wavy diffuse boundary
Bt2	50–80 cm	Yellowish brown (10YR 5/6) clay; moderate fine and medium subangular blocky structures; some medium clay coatings; some moderate mangans (black, 10YR 2/1); moderate firm, very sticky and very plastic; few fine roots; smooth diffuse boundary
Bt3	80–110 cm	Yellowish brown (10YR 5/6) clay; moderate fine and medium subangular blocky structures; some medium clay coatings; many moderate mangans (black, 10YR 2/1); moderate firm, very sticky and very plastic; few very fine roots; smooth diffuse boundary
Bt4	110–150 cm	Yellowish brown (10YR 5/6) clay; moderate fine and medium subangular blocky structures; few thin clay coatings; many moderate mangans (black, 10YR 2/1); moderate firm, very sticky and very plastic; few very fine roots; wavy abrupt boundary

The micromorphological observations by thin sections supported field morphological observations. Soil fabric, the spatial arrangement of the soil particles and voids, varied consistently with horizons, from undifferentiated in the C horizon to unistrial in the Bt horizons. In the C horizon, only

Table 9.1 Physical and chemical properties of pedons YM and PK1

| Horizon | Bd[a] (Mg/m^3) | Clay (%) | Silt (%) | Sand (%) | Sand-sized fraction[b] | | | | | pH | OC[c] (g/kg) | CEC[d] (cmol/kg) | Iron fraction[e] | | |
					VC (%)	C (%)	M (%)	F (%)	VF (%)				Fe$_d$ (g/kg)	Fe$_o$ (g/kg)	Fe$_t$ (g/kg)
Pedon YM															
Ap1	1.5	34	46	20	0	0.6	2.8	9.1	7.5	3.9	13.6	8.8	28.0	5.0	86.7
Ap2	1.5	36	46	18	0	0.5	2.1	8.0	7.0	4.0	10.5	8.6	28.0	5.0	75.3
Bo1	1.7	49	40	11	0	0.4	1.4	5.0	4.4	4.3	7.8	7.4	38.1	6.2	74.2
Bo2	1.7	51	38	11	0	0.3	1.4	4.7	4.0	4.5	3.2	7.4	38.7	6.3	65.8
Bo3	1.7	54	37	9	0	0.4	1.1	4.1	3.7	4.6	2.5	8.7	42.7	6.8	70.1
Bo4	1.7	55	35	10	0	0.4	1.2	4.2	3.7	4.8	2.4	8.5	43.2	7.1	74.8
Bo5	1.7	55	35	10	0	0.3	1.3	4.4	3.9	4.5	2.2	8.2	44.0	6.0	78.3
Pedon PK1															
A	1.6	40	52	8	0	0.2	0.2	1.8	5.7	3.7	14.4	7.3	24.6	1.8	35.8
Bo1	1.5	43	50	7	0	0.1	0.2	1.5	4.9	3.9	4.8	7.0	25.0	1.7	32.3
Bo2	1.5	48	46	6	0	0.1	0.2	1.2	4.1	3.9	3.2	8.4	29.1	2.1	37.8
Bo3	1.5	50	45	5	0	0	0.2	1.3	4.0	3.9	3.6	7.6	28.3	2.2	35.7
Bo4	1.5	48	47	5	0	0	0.2	1.2	4.1	4.0	2.4	7.8	30.0	2.0	34.7
Bo5	1.4	49	46	5	0	0.1	0.1	1.2	4.0	4.1	2.0	7.2	30.4	1.6	33.5

[a] Bulk density

[b] *VC* very coarse (1–2 mm), *C* coarse (0.5–1 mm), *M* medium (0.25–0.5 mm), *F* fine (0.1–0.25 mm), *VF* very fine (0.05–0.1 mm)

[c] Organic carbon

[d] Cation exchange capacity

[e] *d* dithionite-extractable, *o* oxalate-extractable, *t* total content

Copyright by Prof. Hseu

Fig. 9.7 Profile morphology of PK2. Copyright by Prof. Hseu

a few weakly impregnated ferri-mangans were found and no clay coatings occurred. However, the soil of C horizon revealed not only undifferentiated B-fabric with dotted fine materials but also mollusk fossils surround by reef detritus (Fig. 9.10a). Coarse Fe/Mn nodules and clear mangans can be found in the Bt horizons (Fig. 9.10b). Furthermore, clearly illuvial clay coatings can be found in the chamber of soil matrix (Fig. 9.11a), which were strongly well-oriented (Fig. 9.11b). It is interesting to note that the boundaries between Fe/Mn nodules and soil matrix was very clear, indicating the accumulation of secondary Fe and Mn oxides were relict from previous pedogenic processes in wet conditions (Hseu and Chen 1997, 2001).

9.2.4 Physical and Chemical Properties and Clay Mineralogy of Ultisols

Clay contents in the Oxisols and Ultisols are very high, particularly in their Bt horizons with over 40 %. However, the sand contents are almost less than 20 %, which sizes are dominant in the fine and very fine factions. Like above-mentioned Oxisols, the organic carbon content and pH are very low (Table 9.2). The CEC values of the Ultisols are slightly higher than those of the Oxisols. Additionally, the CEC/clay clearly exceed 16 cmol/kg. The Oxisols and Ultisols are highly weathered to produced pedogenic Fe oxides,

Fig. 9.8 Photomicrographs of Bt horizons from Pedon PK2: clay coatings along the pedsurface under plane polarized light (*left*) and cross-polarized light (*right*), the width of frame is 1.0 mm. Copyright by Prof. Hseu

Fig. 9.9 Field morphology of pedon Lu2 on the marine terrace of Liuchiu Island. Copyright by Prof. Hseu

Fig. 9.10 microphotograph of C horizon in Pedon Lu2. Mollusk fossils surround by reef detritus (*left*). Coarse Fe/Mn nodules and clear mangans in the Bt horizon (*right*). The width of photo frame is 0.5 cm. Copyright by Prof. Hseu

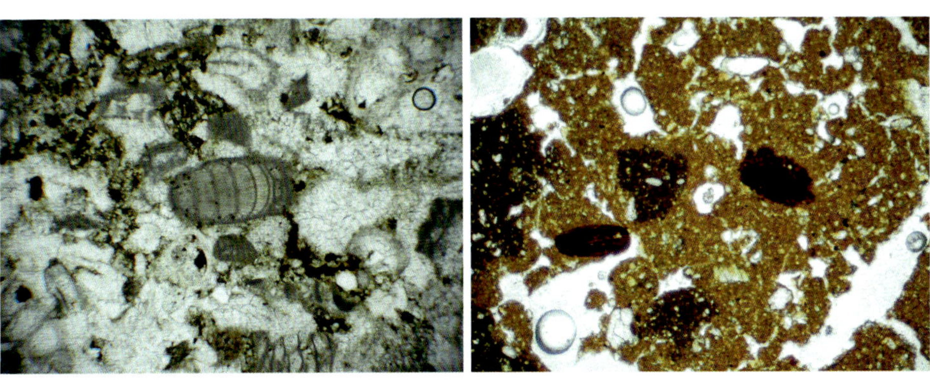

so that the proportion of free Fe (Fe_d, determined by DCB-extraction) in total Fe is very high and the well drainage causes the low proportion of amorphous Fe (Fe_o, determined by oxalate-extraction) in Fe_d, on the terraces.

The particle-size distribution for the soils varied substantially among different levels of marine terrace on Liuchiu Island. However, strong chemical weathering condition characterized by high air temperature and heavy rainfall in the study area has resulted in very low amounts of sand in Pedon Liu2. The Pedon Liu2 have few sand and great amounts of clay (Table 9.3). If the sand fraction was further subdivided into five classes, the dominant classes would be fine sand and very find sand. This indicated that the parent materials of the soils were fine and highly weatherable, and thus, a high rate of soil development can be expected.

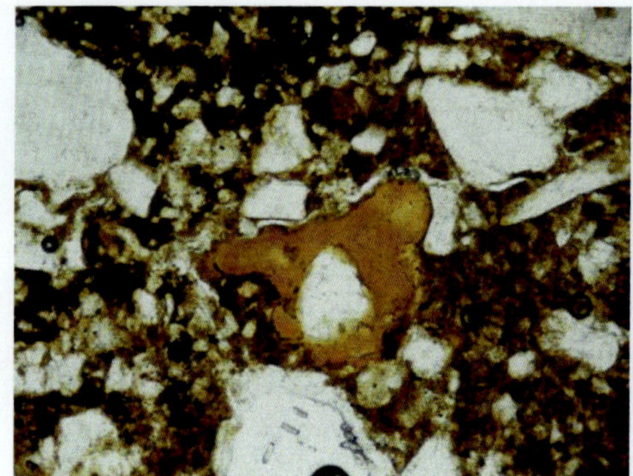

Fig. 9.11 Microphotograph of Bt2 horizon in Pedon Lu2. Illuvial clay coatings in the chamber of soil matrix (*left* plane polarized light, *right* cross polarized light). The width of photo frame is 0.25 cm. Copyright by Prof. Hseu

Table 9.2 Physical and chemical properties of the Ultisol pedons (PK2-5) on the Pakua tableland in Central Taiwan

Pedon	Horizon	Sand	Silt	Clay	Sand-sized fraction[a]					OC^b	pH	CEC^c	BSP^d	CEC/clay	Iron fraction[e]		
					VC	C	M	F	VF						Fe_o	Fe_d	Fe_t
PK-2	A	11	49	40	0	0.2	0.5	2.4	6.3	18.0	3.7	12	6	31	2.2	27.2	36.1
	Bt1	9	47	44	0	0.4	0.5	3.2	5.4	8.2	3.8	9.8	5	22	2.9	26.3	32.4
	Bt2	9	46	45	0.1	0.3	0.5	1.9	4.9	8.4	3.7	9.4	6	21	2.3	24.7	28.2
	Bt3	8	45	47	1.2	0.6	0.6	3.6	5.2	8.0	3.8	8.7	5	19	2.3	27.5	34.5
	Bt4	10	44	46	0	0.3	0.5	3.5	5.8	6.4	3.9	8.0	6	17	2.4	28.7	34.2
	Bt5	10	45	45	0	0.5	0.5	3.6	6.1	6.6	3.9	8.1	5	18	4.1	31.1	37.0
PK-3	A	16	46	38	0.2	0.8	4.3	5.8	5.3	17.2	4.1	10	12	25	4.0	31.1	36.5
	AB	19	42	39	0.4	2.3	2.1	4.4	10.2	15.5	4.1	9.2	13	24	4.2	26.7	38.9
	Bt1	21	48	31	0.2	4.1	2.8	5.3	8.4	8.2	4.2	7.5	12	23	4.5	34.3	36.7
	Bt2	20	39	41	0.3	3.0	5.8	5.3	5.2	7.5	4.2	7.5	11	23	4.3	34.7	36.4
	Bt3	22	40	38	0.3	3.4	2.7	5.8	9.7	7.5	4.3	7.8	8	24	4.5	36.7	39.1
PK-4	A	14	46	40	0.4	2.0	2.2	4.2	5.2	17.5	4.1	13	23	33	2.9	23.8	26.4
	Bt1	15	37	48	0.8	3.1	2.6	4.0	4.1	15.0	4.2	11	13	24	2.6	23.5	31.0
	Bt2	16	35	49	1.1	3.0	3.0	4.9	4.4	7.7	4.2	9.3	14	20	2.8	27.4	33.4
	Bt3	17	35	48	1.9	3.2	2.7	4.9	4.6	7.7	4.3	9.0	17	20	2.8	29.5	35.3
PK-5	A	22	50	28	0.1	1.9	3.1	5.7	11.1	15.0	3.9	9.8	14	39	4.8	31.3	41.3
	AB	19	52	29	0.2	3.0	4.2	5.3	6.3	12.4	4.1	9.4	6	38	5.3	37.9	42.9
	Bt1	19	41	40	0	2.4	3.4	6.4	7.1	7.8	4.1	8.5	6	21	6.6	33.2	45.4
	Bt2	21	41	38	0.4	3.5	3.6	6.3	7.5	5.5	4.2	8.5	9	22	7.1	37.2	43.7
	Bt3	21	36	43	0.5	2.7	3.4	6.1	8.6	4.8	4.2	8.0	7	22	6.3	31.9	42.1

[a] *VC* very coarse (1–2 mm), *C* coarse (0.5–1 mm), *M* medium (0.25–0.5 mm), *F* fine (0.1–0.25 mm), *VF* very fine (0.05–0.1 mm)
[b] Organic carbon
[c] Cation exchange capacity
[d] Base saturation percentage
[e] *d* dithionite-extractable, *o* oxalate-extractable, *t* total content
Copyright by Prof. Hseu

Soil pH shows no consistent trends with depth in Pedon Lu2, but it is always less than 7.0 (Table 9.4). The carbonate was completely leached out and not detectable in Pedon Lu2. The organic carbon content decreased with soil depth and was very low, generally <40 g/kg. However, CEC was low to moderate, ranging from 22.0 to 33.8 cmol/kg. The CEC value was not correlated with pH value, the amount of clay, and the organic matter content. The exchangeable bases showed relatively high amounts of Ca and Mg in all samples, whereas Na and K were much lower. In Soil Taxonomy, Pedon Lu2 is classified as Typic paleudult.

In pedon Lu2, the migration of pedogenic Fe and Mn from the eluvial horizon to the lower parts of profile was indicated by the accumulation of both DCB and oxalate extractable amounts of the two metals in the subsurface horizons (Table 9.5). However, the Al_d concentrations did not increase with depth as uniformly. Compared with DCB-extractable Fe (Fe_d), oxalate-extractable Fe (Fe_o) was much lower so that the

Table 9.3 Physical properties of Pedon Lu2 on the marine terrace of Liuchiu Island

Horizon	Depth (cm)	Sand (%)	Silt (%)	Clay (%)	Sand-sized fraction[a]					Bd[b] (Mg/m^3)
					VC (%)	C (%)	M (%)	F (%)	VF (%)	
A	0–10	5	54	41	0.9	0.8	0.8	0.8	1.7	1.4
AB	10–30	5	47	48	0.1	0.1	0.3	0.4	3.9	1.2
Bt1	30–50	3	40	57	0	0.1	0.1	0.3	2.4	1.4
Bt2	50–80	2	37	61	0	0	0.1	0.2	1.7	1.4
Bt3	80–110	2	35	63	0	0	0	0.1	1.8	1.4
Bt4	110–150	2	31	67	0	0	0	0.1	1.9	1.5
C	>150	5	42	53	0.3	1.5	0.8	1.1	1.3	1.4

[a] *VC* very coarse (1–2 mm); *C* coarse (0.5–1 mm); *M* medium (0.25–0.5 mm); *F* fine (0.1–0.25 mm); *VF* very fine (0.05–0.1 mm)
[b] Bulk density
Copyright by Prof. Hseu

Table 9.4 Chemical properties of Pedon Lu2 on the marine terrace of Liuchiu Island

Horizon	Depth (cm)	pH	Carbonate (%)	OC[a] (g/kg)	CEC[b]	Exchangeable bases				BSP[c] (%)
						K (cmol/kg)	Na (cmol/kg)	Ca (cmol/kg)	Mg (cmol/kg)	
A	0–10	6.5	ND[d]	39	26.6	0.3	0.2	4.6	1.2	23.8
AB	10–30	6.5	ND	26	22.5	0.1	0.2	3.5	1.1	21.9
Bt1	30–50	5.5	ND	11	24.3	0.1	0.2	2.5	1.6	18.3
Bt2	50–80	5.5	ND	7.5	26.9	0.1	0.2	2.6	1.8	17.6
Bt3	80–110	5.8	ND	5.9	27.4	0.1	0.3	2.8	1.9	18.6
Bt4	110–150	6.0	ND	4.3	33.8	0.2	0.3	3.8	1.7	18.0
C	>150	6.2	ND	3.5	22.0	0.2	0.3	4.6	2.0	32.2

[a] Organic carbon
[b] Cation exchange capacity
[c] Bbase saturation percentage
[d] Not detectable
Copyright by Prof. Hseu

Table 9.5 Extraction of Al, Fe and Mn of Pedon Lu2 on the marine terrace of Liuchiu Island

Horizon	Depth (cm)	DCB[a]			Oxalate[b]			Total Fe (g/kg)	Fe$_o$/Fe$_d$	(Fe$_d$–Fe$_o$)/Fe$_t$
		Fe (g/kg)	Al (g/kg)	Mn (mg/kg)	Fe (g/kg)	Al (g/kg)	Mn (mg/kg)			
A	0–10	13.1	1.14	173	2.24	1.64	64.6	20.1	0.17	0.54
AB	10–30	13.6	1.21	157	1.90	1.61	69.5	19.9	0.14	0.59
Bt1	30–50	15.1	1.32	161	2.39	1.74	81.4	29.7	0.16	0.43
Bt2	50–80	15.7	1.28	176	2.38	1.59	89.5	25.6	0.15	0.52
Bt3	80–110	15.2	1.22	195	2.74	1.64	88.5	30.3	0.18	0.41
Bt4	110–150	16.8	1.09	140	2.26	1.61	87.6	42.2	0.13	0.34
C	>150	15.2	1.10	145	1.95	1.58	78.9	48.5	0.13	0.27

[a] Dithionite-citrate-bicarbonate extraction
[b] Oxalate extraction
Copyright by Prof. Hseu

ratio of oxalate and DCB-extractable Fe (Fe$_o$/Fe$_d$) was always less than 0.20, indicating a low degree of pedogenic Fe activity and a well drained condition. Conversely, the crystallinization of the pedogenic Fe was very high based on the profile trend of the values of (Fe$_d$–Fe$_o$)/Fe$_t$ (Fe$_t$ represents total Fe). Unlike Fe$_o$ and Mn$_o$, measurements of oxalate-extractable aluminum (Al$_o$) indicated no trends with depth, because aluminum is not redox-sensitive in soil during pedogenesis (Blume and Schwertmann 1969).

Vermiculite, vermiculite-illite interstratified minerals, illite, kaolinite, quartz, and gibbsite are common in the Oxisols and Ultisols in variable quantities (Table 9.6). Kaolinite and illite are the predominant minerals while chlorite and mectite are absent. Tsai et al. (2006) grouped pedon PK1 (Oxisols) and the other Ultisol pedons (pedons PK2-5) into two sets that largely parallel the terrace altitude from which the pedons were collected. Pedons PK1 and PK2 contain higher amounts of kaolinite and gibbsite but lower amount of randomly interstratified minerals, and pedons PK4 and PK5 contain lower kaolinite and gibbsite but higher randomly interstratified minerals. Pedons of the first set are from the terraces at higher altitude, the second set are from the terraces at lower altitude. The presence of kaolinite and gibbsite as well as the absence of montmorillonite indicate high degrees of leaching and desilication of the soils (Arduino et al. 1986), resulting in the decrease in weatherable primary minerals content. In comparison, the interstratified minerals of vermiculite-illite as product of alteration from illite represent a weaker degree of leaching and desilication of the soils. The soils of the first set seem to have undergone stronger or longer weathering in the past, and can be regarded as older in age than the others. This confirms that the terraces at higher altitude are older than those at the lower altitude.

9.3 Ultisol with Redoximorphic Features

Ultisols are mainly found in the paddy soils of red earth and aged alluvium from the red earth on Quaternary terraces of northwestern part of Taiwan. These soils have been used for lowland rice production for over 50 years, and the significant morphological characteristics of the surface epipedon were the color change owing to flooding by irrigation water. The diagnostic subsurface horizon is argillic horizon with plinthites that are iron-rich, humus poor mixtures of clay with quartz and other minerals (Soil Survey Staff 2010). It commonly occurs as dark red redox concentrations that usually form platy, polygonal or reticulate patterns (Fig. 9.12). It changes irreversibly to an ironstone hardpan or to irregular aggregates on exposure to repeated wetting and drying, especially if it is also exposed to heat from the sun. Therefore, the Subgroups of plinthitic Ultisols are Plinthic Paleudults and Typic Plinthaquults, respectively.

In general, plinthitic soils are intermittently saturated with water (Vepraskas and Wilding 1983; Jacobs et al. 2002). It has also been accepted that a fluctuating water table is necessary to concentrate Fe or Mn in plinthitic zones (Perkins and Lawrence 1982). Soft masses of Fe and ferromanganiferous nodules are formed in these plinthitic soils with strong segregation of Fe and Mn. It is very important to understand the formation of ferromanganiferous nodules for predicting regional hydrology and evaluating soil degradation within a landscape unit (Bouma 1983; Vepraskas 1994; Stiles et al. 2001; Aide et al. 2004).

9.3.1 Morphology of Plinthitic Ultisols

The Chuwei series in Taoyuan, Typic Plinthaquult, is developed on Quaternary alluvial deposits which have a minimum thickness of 5 m. The soil in the area is commonly utilized for agricultural production, generally in a double-cropping paddy system, and is drained artificially. In this area, the first rice-planting begins in March and is harvested in July, and the second rice-planting begins in September and is harvested in December. Because the cultivation of lowland rice forms a compacted subsurface plowed layer in the soils, the irrigated water is about 10–15 cm above the soil surface while rice is planted during the growing season. The mean air temperature is 27 °C in summer and 13 °C in winter. Mean annual rainfall is approximate 1,560 mm. Moreover, monthly rainfall in winter was considerably less than in other seasons during the past several decades. The wet season is from February to June, indicating that rainfall exceeds evaporation. The field morphological characteristics of Chuwei series on Quaternary alluvial deposits in Taoyuan, Northern Taiwan are described as follows (Fig. 9.2):

On the red earth terrace of Northern Taiwan, Jien et al. (2010) further indicated that the greatest quantity of ferromanganiferous nodules (91–492 g/kg) in the Ultisols with

Ap	0–20 cm	Brown (10YR 4/3) ped interiors with common (5 %), light yellowish brown (2.5Y 6/4) fine mottles, and few (10 %) brownish yellow (10YR 6/8) very fine and fine mottles; common (10 %), prominent, fine and medium, red (2.5YR 4/8) irregular Fe nodules in the matrix; silty clay loam; moderate very fine granular parting to moderate very fine and fine angular blocky; slightly sticky and slightly plastic; gradual smooth boundary
AB	20–36 cm	Very dark grayish brown (2.5Y 3/2) ped interior with common (10 %), very fine reddish yellow (7.5YR 6/8) mottles, and common (5 %), fine yellowish brown (10YR 5/4) mottles; common (15 %), prominent, fine and medium red (2.5YR 4/6) Fe nodules, irregular in the matrix; clay loam; moderate very fine, fine, and medium angular blocky; slightly sticky and slightly plastic; clear wavy boundary
Btv1	36–70 cm	Light gray (2.5Y 7/1) ped interior with many (20 %) prominent 5YR 4/8 medium and coarse irregular mottles in the matrix; common (10 %) prominent yellowish brown (10YR 5/4) fine and medium irregular mottles in the matrix; common (15 %) prominent fine and medium red (2.5YR 4/6) irregular Fe nodules in the matrix; clay; massive; sticky and strong plastic (40–50 % continuous plinthites); diffuse boundary

(continued)

(continued)

Btv2	70–110 cm	Light gray (2.5Y 7/1) ped interior with many (35 %) distinct medium and coarse red (2.5YR 4/8) irregular mottles in matrix; and common (20 %) prominent yellowish brown (10YR 5/6) medium and coarse irregular mottles in the matrix; common (10 %) prominent dark red (2.5YR 3/6) fine and medium irregular Fe nodules in the matrix and few (2 %) black (2.5Y 2.5/1) very fine and fine Mn masses; clay; massive; sticky and strong plastic; diffuse boundary; continuous plinthites
Btv3	110–145 cm	Light gray (2.5Y 7/1) ped interior with many (50 %) prominent yellowish brown (10YR 5/8) coarse and very coarse irregular mottles in matrix; common (5 %) prominent red (2.5YR 4/8) (5 %) fine and medium irregular mottles in the matrix, and common (5 %) black (2.5Y 2.5/1) fine Mn nodules; clay; massive; strong sticky and strong plastic; diffuse boundary; continuous plinthites
Btv4	145–170 cm	Light gray (2.5Y 7/1) ped interior with common (20 %) prominent red (2.5YR 5/8) coarse and very coarse irregular grey mottles in matrix and many (30 %) prominent brownish yellow (10YR 6/8) coarse mottles; clay; massive; strong sticky and strong plastic; gradual wave boundary; continuous plinthites
Btv5	>170 cm	Light gray (2.5Y 7/2) ped interior with common (5 %) prominent yellowish red (5YR 5/8) fine and medium irregular mottles in matrix and many (40 %) prominent reddish yellow (7.5YR 6/8) medium mottles; silty clay loam; massive; sticky and slightly plastic

Fig. 9.12 Field morphology of plinthites in Taoyuan, Northern Taiwan (The number unit on the ruler is cm). Copyright by Prof. Hseu

various redoximorphic features were found in plinthitic horizons which had reducing conditions for a moderate amount of time (about 47 % of the year). This was especially true for nodules >20 mm in diameter. Therefore, a moderate duration of reducing conditions is suitable for formation and growth of the nodules. Micromorphological observation of ferromanganiferous nodules revealed that co-presentation of secondary Fe and Mn oxides and oriented clay is more apparent in coarser nodules than finer nodules (Fig. 9.13). These observations evidence that the formation of nodules was impregnated with clay and continued accumulation of pedogenic Fe and Mn. It was deduced that ferromanganiferous nodules were formed when Mn (II) first precipitated within micropores or coated the coarse nuclear grain surface. Later, reduced Fe, Mn and illuvial clay continued reoxidizing and precipitating on the initial Mn nodules and then grew to a coarser size. Iron activity ratio (Fe_o/Fe_d) is suggested as a good index for prediction of annual duration of reducing conditions in the soils examined in this study. Furthermore,

the ratio of Mn_d/Fe_d in nodules is a good indicator for prediction of soil depths where the water table fluctuated, because Mn accumulated and was well crystallized at these depths.

The Chuwei soil has very high clay content in any horizon (Table 9.7), except for its AB horizon. Clay illuviation is proved by clear clay coating in the Btv horizons. For each fraction of sand, medium and find sand fractions are dominant through all horizons, indicating homogenous soil horizon of the Chuwei soil. Some anthropic disturbances must have been occurred in surface soils because textural discontinuity was found between Ap and AB horizons, especially for clay contents and coarse and medium sand fractions. The Chuwei soil was characterized by a pronounced acidic reaction (pH $\leq$ 5.5) and very low soil organic carbon (SOC) concentration ($\leq$1.0 %), which could be resulted from rotation cropping of watermelon and rice. The CEC values for the Chuwei soil ranged from 7.40 to 13.0 cmol/kg and increased with clay and SOC contents in these horizons (Table 9.8). High base saturation in the soils could result from long-term fertilizer application.

Similar distribution through the Chuwei profile was found among Fe, Al, and Mn with the DCB extraction (Table 9.9). Additionally, Fe_d, Al_d and Mn_d contents increased with

Table 9.6 Clay mineral composition of the Oxisol and Ultisol pedons on the Pakua tableland in Central Taiwan

Pedon	Horizon	G	K	V	V–I	I	Q
Oxisol							
PK-1	A1	+++	+++	+	+	++	++
Ultisol							
PK-2	A	+++	+++	+	+	++	++
	Bt1	++	+++	+	+	+++	++
	Bt2	++	++	+	+	+++	++
	Bt3	++	++	+	+	+++	+
	Bt4	+++	++	+	+	+++	+
	Bt5	+++	+++	+	+	++	+
PK-3	A	+++	+++	+	+	++	++
	AB	++	+++	+	+	+++	+
	Bt1	++	++	+	+	++	+
	Bt2	++	++	+	+	++	+
	Bt3	++	++	+	++	++	+
PK-4	A	++	+++	+	+	+++	+
	Bt1	++	++	+	++	+++	+
	Bt2	++	++	+	++	++	+
	Bt3	++	++	+	++	++	+
PK-5	A	++	+++	+	+	++	+
	AB	++	++	+	++	+++	+
	Bt1	++	++	+	+	++	+
	Bt2	++	++	+	++	++	+

I illite, *K* kaolinite, *V* vermiculite, *Q* quartz, *G* gibbsite, *V–I* interstratified minerals of vermicular-illite

+++ 25~50 %, ++ 10~25 %, + <10 %

Copyright by Tsai et al. (2006)

increasing of clay contents within soil depth, suggesting co-illuviation of Fe–Mn oxides and clay. The Chuwei soil had the clear migration of Fe_d; moreover, the extremely high Fe_d/Fe_t ratio (≥ 0.90), as the lowest Fe_o/Fe_d ratio, was found in this soil, which indicates that the Chuwei soil experienced the strongest weathering. Regarding Fe_o, the highest contents of Fe_o were found in surface soils (Ap and AB horizons), which indicated that soil organic matter (SOM) influenced the crystallization of Fe oxides. Additionally, the Fe_o/Fe_d ratio was always decreased with depth along the profile, which could be attributed to higher SOM content in surface soils that hampered crystallization of Fe (Schwertmann and Fanning 1976). Brahy et al. (2000) studied the soils along a transect in a forest in Belgium and found that the Fe_o/Fe_d ratio was higher in the soil surface and Cambisol soils, which resulted from higher amounts of SOM in surface soils and hydromorphic environments in soils. However, very high Fe_o content was also found in the Btv5 horizon (>170 cm) of Chuwei soil where exists frequent fluctuation of ground water table (Jien et al. 2010), and frequent saturation and reduction conditions in this horizon might also retard crystallization of Fe/Mn oxides.

The distribution of Mn is similar with Fe within the Chuwei soil, two sudden increasing concentrations of Mn_o, Mn_d and Mn_t were found in the Chuwei pedon. The first concentration peak was found in Btv1 horizon where Fe and Mn illuviated from upper reduced soil horizon and then accumulated here. The second peak was found in Btv5 horizon and it could be resulted from illuviation and accumulation of reduced Mn from Btv3, Btv4, and Btv5 horizons with following oxidization of Fe and Mn in this horizon. The most quantities of ferromanganiferous nodules found in the Btv5 horizon could further support the accumulation phenomenon of Fe and Mn we mentioned above.

9.4 Formation and Genesis

Quaternary terrace of aged alluvium is the most typical landscape with Oxisols and Ultisols in Taiwan. The strata under terrace gravels are composed of coarse-conglomerate deposits of the upper Pleistocene Tokoshan formation (0.8–0.5 Ma) (Chen and Liu 1991). These conglomerates are characteristically well sorted with consolidated cementation, and sometimes contain fractures due to prolonged tectonic compression

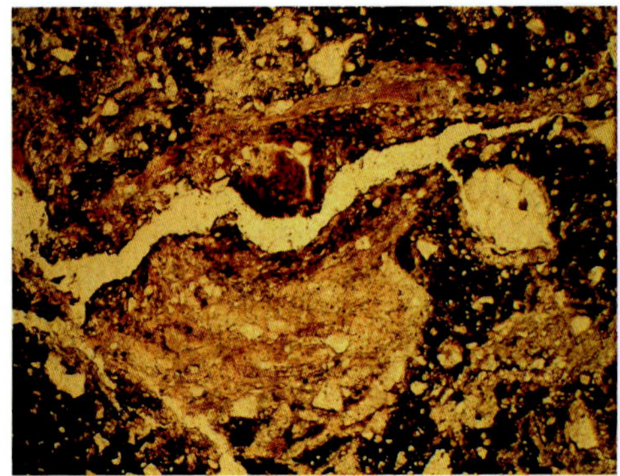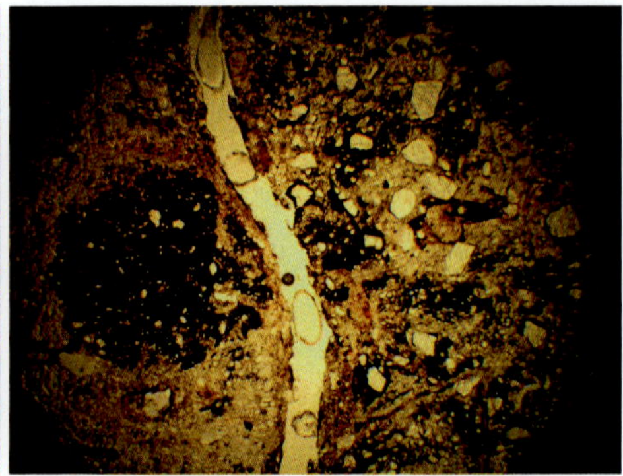

Fig. 9.13 Micrographs of redoximorphic features in the Btv3 horizon of Chuwei series in Taoyuan (Typic Plinthaquult). *Left* co-presentation of secondary Fe and Mn oxides and oriented clay; *Right* ferromanganiferous nodules. Copyright by Prof. Hseu

Table 9.7 The particle size distribution of Chuwei series

Horizon	Depth (cm)	Clay (%)	Silt (%)	Sand (%)	Sand fraction[a]					Bd[b] (%)
					VC (%)	C (%)	M (%)	F (%)	VF (%)	
Ap	0–20	60.7	24.3	15.0	0.5	0.6	2.9	5.6	5.5	1.55
AB	20–36	36.1	48.3	15.6	0.6	2.1	8.6	2.8	1.4	1.56
Btv1	36–70	58.5	32.3	9.2	0.3	0.5	1.6	4.6	2.1	1.63
Btv2	70–110	74.9	17.2	7.9	0.0	0.5	1.7	3.4	2.1	1.53
Btv3	110–145	69.4	19.7	10.9	0.0	0.3	2.4	5.9	2.0	1.65
Btv4	145–170	70.6	17.8	11.6	0.0	0.2	3.9	5.4	1.9	1.65
Btv5	>170	73.1	15.8	11.2	0.9	0.9	3.2	4.1	1.7	1.68

[a] VC very coarse (2–1 mm), C coarse (1–0.5 mm), M medium (0.5–0.25 mm), F fine (0.25–0.1 mm), VF very fine (0.1–0.05 mm)
[b] Bulk density (Mg/m^3)
Copyright by Prof. Hseu

Table 9.8 The selected chemical properties of Chuwei series

Horizon	Depth (cm)	pH		O.C.[a] (g/kg)	Exchange bases					BSP[c] (cmol (+)/kg)
		H_2O	KCl		CEC[b] (cmol (+)/kg)	Na (cmol (+)/kg)	K (cmol (+)/kg)	Ca (cmol (+)/kg)	Mg (cmol (+)/kg)	
Ap	0–20	5.0	4.4	9.9	7.4	0.70	0.40	3.1	0.7	66
AB	20–36	5.3	4.7	10	7.5	0.70	0.50	3.3	1.8	84
Btv1	36–70	5.8	4.8	3.2	9.2	0.30	0.60	3.2	2.7	74
Btv2	70–110	6.1	5.0	2.1	11	0.20	0.50	3.1	3.2	64
Btv3	110–145	6.2	4.9	1.8	9.3	0.20	0.40	2.6	0.7	42
Btv4	145–170	6.2	5.0	1.3	9.9	0.20	0.40	3.5	1.8	60
Btv5	>170	6.2	5.0	1.2	13	0.20	0.50	3.9	2.4	54

[a] Organic carbon
[b] Cation exchangeable capacity
[c] Base saturation percentage
Copyright by Prof. Hseu

Table 9.9 The selected extraction of Fe, Al, and Mn of Chuwei series

Horizon	Depth (cm)	Fe_d^a (g/kg)	Fe_o^b (g/kg)	Fe_t^c (g/kg)	Al_d^a (g/kg)	Al_o^b (g/kg)	Mn_d^a (g/kg)	Mn_o^b (g/kg)	Mn_t^c (g/kg)
Ap	0–18	22 ± 2.1	2.6 ± 0.0	22 ± 4.0	3.5 ± 0.5	1.1 ± 0.0	0.1 ± 0.0	0.1 ± 0.0	0.2 ± 0.0
AB	18–36	26 ± 0.3	2.5 ± 0.0	26 ± 2.0	3.8 ± 0.4	1.1 ± 0.0	0.1 ± 0.0	0.1 ± 0.0	0.2 ± 0.0
Btv1	36–70	32 ± 3.1	1.7 ± 0.1	39 ± 2.0	4.0 ± 0.1	1.3 ± 0.0	0.4 ± 0.1	0.4 ± 0.0	0.6 ± 0.0
Btv2	70–110	37 ± 0.7	1.5 ± 0.0	34 ± 10	4.9 ± 0.2	1.4 ± 0.1	0.2 ± 0.0	0.2 ± 0.2	0.3 ± 0.2
Btv3	110–145	22 ± 1.4	0.9 ± 0.1	23 ± 0.7	3.8 ± 0.4	1.0 ± 0.0	0.2 ± 0.1	0.3 ± 0.1	0.3 ± 0.1
Btv4	145–170	27 ± 0.1	0.5 ± 0.4	29 ± 2.0	2.7 ± 0.5	0.5 ± 0.0	0.0 ± 0.0	0.0 ± 0.2	0.1 ± 0.3
Btv5	>170	60 ± 0.4	2.6 ± 0.1	66 ± 1.3	4.5 ± 0.1	0.7 ± 0.0	6.9 ± 0.1	7.2 ± 0.0	9.2 ± 0.0

[a] Free iron, manganese, and aluminum oxides
[b] Amorphous iron, manganese, and aluminum oxides
[c] Total amounts of iron, manganese, and aluminum in soils
Copyright by Prof. Hseu

of mountain building during the Plio-Pleistocene (Ho 1988). However, the boundary between terrace gravels and the conglomerate deposits is often ambiguous in the field (Tsai and Sung 2003). The parent materials consist of quartzite, sandstone, and shale, with minor slate.

In the Ultisols for paddy rice production of Taiwan, saturation and reduction are the common soil characteristics, and various redoximorphic features occur with anthraquic or oxyaquic conditions (Hseu and Chen 1994, 1996, 1997, 2001; Jien et al. 2010). The concept of aquic conditions was introduced in the *Keys to Soil Taxonomy* to assess seasonal wetness throughout the soil profile rather than just using indicators within the upper 50 cm of the pedon for moisture category (Soil Survey Staff 1992). Alternating cycles of reduction and oxidation in soils over prolonged periods, and the consequent mobility and accumulation or depletion of Fe and Mn, result in the formation of redoximorphic features (Vepraskas 1994). Seasonally irrigated flooding and fluctuation of groundwater level control the redoximorphic features, saturation, and reduction of rice-growing Ultisols in different landscape positions of Taiwan. Past studies have tested the application of redoximorphic features as soil saturation indicators in various pedogenic environments.

Hseu and Chen (2001) have selected three paddy soils from Quaternary terrace along a toposequence of the Chungli Terrace in Northern Taiwan for monitoring of water table, matric potential, and redox potential (Eh) at various soil depths in 1996 and 1997. The three soils are Houhu (Typic Plinthaquult) in the toeslope, Hsinwu (Typic Plinthaquult) in the footslope, and Lungchung (Plinthaquic Paleudult) in the lower backslope. Redox concentrations originally occurred as soft masses and concrete nodules associated with seasonally high water levels, but irrigation and drainage processes also influenced the development of redoximorphic features. The studied Fe–Mn concretions and Fe depletions increased with increasing the cycling of oxidation and reduction in rice production. The durations of saturation and reduction in the Btv horizons (argillic horizons with plinthites) of the Houhu soil in the toeslope position were more than 80 % of the year and the soil had about 10 % of Fe–Mn concretions. The Btv horizons of the Hsinwu soil in the footslope position were saturated for 50 % of the year and reduced for 25 % of the year, and the soil had about 20 % of Fe–Mn concretions. The Btv horizons of the Lungchung soil in the lower backslope position were saturated for 40 % of the year and reduced for only about 10 % of the year, and the soil had 15 % of Fe–Mn concretions. The Houhu and Hsinwu soils were belonging to the anthraquic condition and the Lungchung soil with less reduction was proposed as having oxyaquic conditions as defined in U.S. soil taxonomy. However, the optimum durations of saturation and reduction were about 50 % of the year in the formation of redoximorphic features within the landscape unit, and thus indicating that semi-quantitative soil hydrology can be estimated by the above redoximorphic features for the paddy soils.

Regarding the Ultisols on coral-reef terrace, Muhs et al. (1987) summarized four possible parental materials for the Terra Rossa-like soils as following: (1) the insoluble residuum of the carbonate limestone; (2) the alluvial sediments from topographic higher positions, stuffing up lower karstified depression; (3) the volcanic ash from neighboring volcano fallen above the carbonate surface; and (4) the eolian dust transported from distant arid regions. Terra Rossa-like soils, found on the marine terraces associated with coral reef in southern part of main Taiwan Island and its off-shore islands (Liu 1992; Cheng et al. 2011), follow these characteristics of parent materials and develop the high amounts of clay and base cations in the Alfisols.

9.5   Land Use and Management

Land use type is very diverse on the Ultisols of Quaternary terrace of aged alluvium, because the landscape is flat and the soil quality is high for agricultural production. However, high clay content, low pH, and high concentrations of free oxides of Fe, Al, Mn are probably major limits in cropping. Since twentieth century, many irrigation systems have been constructed to supply the water use in agriculture, these soils became very productive in rice and other upland crops like pineapple, tea, watermelon, and litchi.

The area of these soils for agricultural is clearly decreasing, because the increase of population and industry requirement. Many industrial parks have been set up on these areas. The industrial activities negatively affect the soil quality on these terraces. Many contaminated sites with heavy metals have been identified on these soils, because the contaminants was from the illegal discharge of industrial wastewater. In this chapter, the soil contamination and remediation will be introduced detailed. In addition, the industrial production is getting to wrest water from the agricultural compartment. The cover of housing and highway on these soils is much higher than the other soils on the slopeland. However, water erosion is clear in these soils because of their terrace shape. The most urgent strategy for conserving these soils is to reduce the erosion and maintain water balance.

References

Aide M, Pavich Z, Lilly ME, Thornton R, Kingery W (2004) Plinthite formation in the coastal plain region of Mississippi. Soil Sci 169:613–623

Arduino E, Barberis E, Marson Ajmone F, Zanini E, Franchini M (1986) Iron oxides and clay minerals within profiles as indicators of soil age in northern Italy. Geoderma 37:45–55

Blume HP, Schwertmann U (1969) Genetic evaluation of profile distribution of aluminum, iron, and manganese oxides. Soil Sci Soc Am Proc 33:438–444

Bouma J (1983) Hydrology and soil genesis of soils with aquic moisture regime. In: Wilding LP, Smeck NE, Hall GF (eds) Pedogenesis and soil taxonomy. Elsevier, Amsterdam, pp 253–281

Brahy V, Deckers J, Delvaux B (2000) Estimation of soil weathering stage and acid neutralizing capacity in a toposequence Luvisol-Cambisol on loess under deciduous forest in Belgium. Eur J Soil Sci 51:1–13

Chen YG, Liu TK (1991) Radiocarbon dates of river terraces along the lower Tahanchi, Northern Taiwan: their tectonic and geomorphic implications. J Geol Soc China 34:337–347

Cheng CH, Jien SH, Tsai H, Hseu ZY (2011) Geomorphological and paleoclimatic implications of soil development from siliceous materials on the coral-reef terraces of Liuchiu Island in Southern Taiwan. Soil Sci Plant Nutri 57:114–127

Guo HY, Hseu ZY, Chen ZS, Tsai CC, Jien SH (2009) Integrating soil information system with agro-environmental application in Taiwan. In: Suh JS, Kim KH, Yoo JH, Kim JK, Paik MK Kim WI (eds) Proceedings of the 9th international conference of east and southeast Asia federation of soil science societies (ESSAFS), pp 117–122. Published by Korean Society of Soil Science and Fertilizer, Seoul, 27–30 Oct 2009. ISBN: 978-89.480-03901 (93520)

Ho CS (1988) An introduction to the geology of Taiwan: explanatory text of the geologic map of Taiwan, 2nd edn. Central Geological Survey of Taiwan, 165 pp

Hseu ZY, Chen ZS (1994) Micromorphology of rice-growing Alfisols with different wetness conditions in Taiwan. J Chinese Agri Chem Soc 32:647–656

Hseu ZY, Chen ZS (1996) Saturation, reduction, and redox morphology of seasonally flooded Alfisols in Taiwan. Soil Sci Soc Am J 60:941–949

Hseu ZY, Chen ZS (1997) Soil hydrology and micromorphology of illuvial clay in an Ultisol hydrosequence. J Chinese Agric Chem Soc 35:503–512

Hseu ZY, Chen ZS (2001) Quantifying soil hydromorphology of a rice-growing Ultisol toposequence in Taiwan. Soil Sci Soc Am J 65:270–278

Jien SH, Hseu ZY, Chen ZS (2010) Hydropedological implications of ferromanganiferous nodules in rice-growing plinthitic Ultisols under different moisture regimes. Soil Sci Soc Am J 74:880–891

Jocobs PM, West LT, Shaw JN (2002) Redoximorphic features as indicators of seasonal saturation, Lowndes County, Georgia. Soil Sci Soc Am J 66:315–323

Liao JH, Wang HH, Tsai CC, Hseu ZY (2006) Litter production, decomposition and nutrient return of uplifted coral reef tropical forest. Forest Ecol Manage 235:174–185

Liu JC (1992) Morphology, pedogenic processes and classification of soils in Kenting region, Taiwan, MS Thesis (in Chinese, with English abstract)

Macleod DA (1980) The origin of the red Mediterranean soils in Epirus, Greece. J Soil Sci 31:125–136

Muhs DR, Crittenden RC, Rosholt JN, Bush CA, Stewart K (1987) Genesis of marine terrace soils, Barbados, West Indies: evidence from mineralogy and geochemistry. Earth Surf Proc Land 12:605–618

Perkins HP, Lawrence B (1982) Sesquioxide segregation in plinthic and nonplinthic counterpart soils. Soil Sci 133:314–318

Schwertmann U, Fanning DS (1976) Iron and manganese concretions in hydrosequences of soils in loess in Bavaria. Soil Sci Soc Am J 40:731–738

Shih TT, Chang JC, Hsu MY, Shen SM (1991) The marine terrace and coral reef dating in Liuchiu Yu, Taiwan. Geogr Res 17:85–97 (in Chinese with English abstract)

Soil Survey Staff (1992) Keys to soil taxonomy. Soil management support service tech. monographs no. 19, 5th edn. Pocahontas Press, Blacksburg

Soil Survey Staff (2010) Keys to soil taxonomy, 119th edn. USDA-NRCS, Washington

Stiles CA, More CI, Driese SG (2001) Pedogenic iron-manganese nodules in Vertisols: a new proxy for paleprecipitation? Geology 29:943–946

Tsai CC, Chen ZS (2000) Lithologic discontinuity of Ultisols along a toposequence in Taiwan. Soil Sci 165:587–596

Tsai H, Hseu ZY, Huang WS, Chen ZS (2007) Pedogenic approach to the geomorphic study on river terraces of the Pakua tableland in Taiwan. Geomorphology 83:14–28

Tsai H, Huang WS, Hseu ZY, Chen ZS (2006) A river terrace soil chronosequence of the Pakua tableland in central Taiwan. Soil Sci 171:167–179

Tsai H, Sung QC (2003) Geomorphic evidence for an active pop-up zone associated with the Chelungpu fault in central Taiwan. Geomorphology 56:31–47

Tsai H, Maejima Y, Hseu ZY (2008) Meteoric [10]Be dating of highly weathered soils from fluvial terraces in Taiwan. Quat Intern 188:185–196

Tsui CC, Hseih CF, Chen ZS (2004) Relationship between soil properties and landscape in a lowland rain forest of Southern Taiwan. Geoderma 123:131–142

Vepraskas MJ (1994) Redoximorphic features for identifying aquic conditions. Tech Bull 301. North Carolina Agric Res Serv, North Carolina State University, Raleigh

Vepraskas MJ, Wilding LP (1983) Aquic moisture regimes in soils with and without low chroma colors. Soil Sci Soc Am J 47:280–285

Yaalon DH (1997) Soils in the Mediterranean region: what makes them different? Catena 28:157–169

Yang GS (1986) A geomorphological study of active faults in Taiwan—especially on the relation between active faults and geomorphic surfaces. Ph.D dissertation, National Normal University, Taipei, 178 pp (in Chinese with English abstract)

10.1 Soil Degradation in Taiwan

The climate of Taiwan is characterized by high air temperature and heavy annual rainfall in association with complicated parent materials of soils. The agricultural soils are intensively tilled for crop production to be satisfied with the large population (about 23 million) in Taiwan. The soil degradation impacted from the above natural conditions and human activities are therefore seen. The area and classification of degraded soils in Taiwan are shown in Table 10.1. The total area of cultivated soils in Taiwan is about 880,000 ha (24 % of the total area), while over 80 % of the cultivated soils are potentially degraded. The main degraded and contaminated soils can be listed as follows (Chen et al. 1996; Chen 1998).

10.1.1 Soil Erosion

The problems of soil erosion in Taiwan are mainly driven by water, and to a less extent by wind. Careless human activities, coastal sand dune, young river terrace, and mudstone bad-land, and slope land are strongly eroded landscapes (Fig. 10.1). However, if these lands are illegally cultivated with vegetables or orchards, the strength of soil erosion will be accelerated due to the reduction of vegetative cover from forests (Fig. 10.2).

10.1.2 Strong Acid Soils

Acid rain and chemical fertilization are major factors that cause soil acidification (pH < 5.6) in Taiwan. Approximately 70 % of the monitored rainfall events by 20 metrological stations through Taiwan have a pH lower than 5.6 in rainwater, respectively. In general, more than 50 % of the acid rain in Taiwan was significantly affected by northeastern flow and frontal passage in the dominant precipitation

systems. Therefore, the chemical composition of rainwater in Taiwan is mainly affected by external materials from other regions, especially from Mainland China. Moreover, serious acid rain in Taiwan often occurs in the highly urbanized and industrialized areas with mean pH values of 4.5. Chen et al. (1998a, b) investigated two heavy metal contaminated soils of Taiwan treated with simulated acid rain (pH 4.0) by crop pot experiments. They found that the simulated acid rain significantly increased the accumulation of Cd, Cu, Pb, and Zn in brown rice and leaves of pickled cabbage based on the annual rainfall of 1,800 mm. Regarding chemical fertilization, approximately 500 million tons of chemical fertilizers of nitrogen, phosphorus, and potassium are applied annually on agricultural soils in Taiwan, which causes a clear increase in acidity in these soils under strong intensities of cropping due to the replacement of base cations with hydrogen ions.

10.1.3 Poorly Drained Soils

Oxygen deficiency and organic acid accumulation in the rhizosphere of crop commonly result from poor drainage of the soils, which reduces metabolic activities and uptake of nutrient and water. Moreover, the operation of cropping machine becomes difficult because of poor drainage in the soil. Poor drainage is generally defined for upland soils in Taiwan. However, some paddy soils for lowland rice production are grouped as poor drained soils if their percolation rates are too low in association with clear organic acid production to be rotated for upland crops in winter. The major reason for poor drainage in Taiwan is the high water table or a large amount of clay in the soils. Additionally, poor-drained soils are also commonly found on the alluvial plains in Taoyuan, Hsinchu, and Ilan counties where annual rainfall is higher than in other regions of Taiwan. The area of poor-drained soil is about 42,300 ha, while the area of very poor-drained soil is about 5,400 ha.

Z.-S. Chen et al., *The Soils of Taiwan*, World Soils Book Series,
DOI 10.1007/978-94-017-9726-9_10, © Springer Science+Business Media Dordrecht 2015

Table 10.1 The main degraded soils in Taiwan

Type of soil degradation	Area (×1,000 ha)	Geographic distribution	Soil classification	
			Soil taxonomy[a]	WRB[b]
Eroded slopeland soils	220	Central Ridge	Entisols	Leptosols
			Inceptisols	Cambisols
Strongly acid soils	250	Northern Taiwan	Ultisols	Alisols
			Oxisols	Acrisols
Poorly-drained soils	180	Northern and Southern alluvial plains	Inceptisols	Cambisols and Gleysols
			Entisols	Leptosols
Highly compact Taiwan clay soils	20	Southern Taiwan	Inceptisols	Cambisols
Salt-affected soils	25	Western coast	Inceptisols	Solonchak
			Entisols	Solonchak
Deficiency in micronutrients	5	Eastern Taiwan	Mollisols	Chernozem
			Entisols	Leptosols
Water stress in deep sandy soils	40	Western coast	Entisols	Fluvisols
Heavy metal contaminated soils	1	Northern and central Taiwan	Ultisols	Alisols
			Entisols	Fluvisols
Total area	741			

[a] Soil Survey Staff (2010)
[b] IUSS Working Group (2006)
Copyright by Profs. Hseu and Chen

Fig. 10.1 Common eroded landscapes in Taiwan: **a** wind erosion on aeolian soil, **b** river terraces from aged alluvial materials, **c** sand dune along coast region, and **d** water erosion on mudstone bad-land. Copyright by Profs. Hseu and Chen

10.1.4 Highly Compacted Taiwan Clay Soils

After long-term and continuous tillage on the cropped lands, the porosity of subsurface soil becomes very low to form firm and platy soil structures, poor aeration and drainage, and difficulty in root penetration, particularly in clayey soils. These compacted soils with long-term tillage for lowland rice production (over about 50 years) are predominantly

Fig. 10.2 Accelerated soil erosion by human activities on **a** slope land, **b** mudstone badland, **c** river terraces from aged alluvial materials, and **d** riparian land. Copyright by Profs. Hseu and Chen

found in some clayey soils from aged lacustrine deposits on the plains of southwestern Taiwan, commonly called by local farmers as "Taiwan clay soils". In general cases of paddy soils, these compact layers are found in depths from 30 to 50 cm (Fig. 10.3). The soil compaction in the paddy field is also one of the reasons for poor drainage. Therefore, if paddy soils are reclaimed for upland crop, the compacted layers need to be destructed for preventing from toxicity of organic acid accumulation.

10.1.5 Salt-Affected Soils

Land subsidence by overpumping of groundwater for freshwater aquaculture locally occurs along the coastal areas of western Taiwan, and thus the coastal soils are usually flooded with seawater, particularly in summer. Thereafter, the soils become salt-affected. Additionally, over-fertilization by livestock compost and chemical fertilizer is another cause for saline accumulation in the soils (Fig. 10.4). The total area of salt-affected soils (>4.0 dS/m in electrical conductivity with saturated soil paste extract) is about 25,000 ha.

10.1.6 Deficiency in Micronutrients

In all trace elements for essential nutrition of crop growth, boron, zinc, iron, and manganese have been considered as

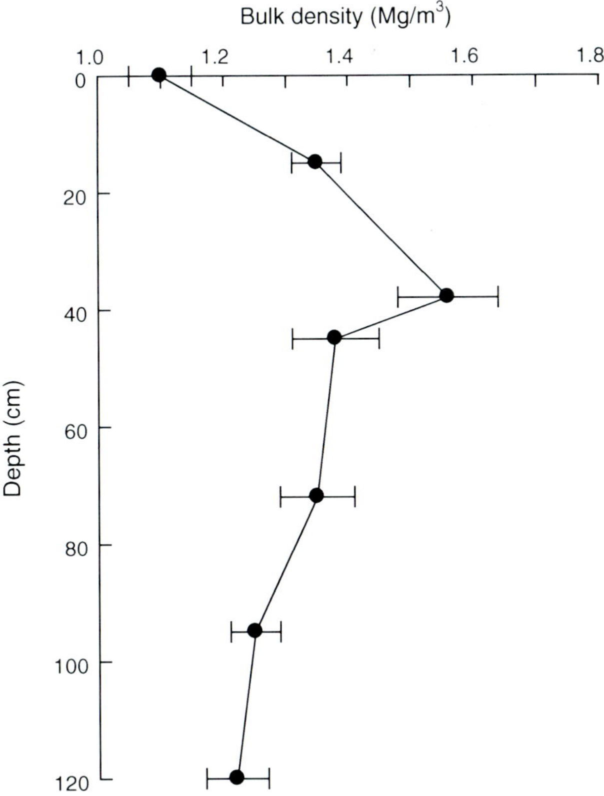

Fig. 10.3 Vertical distribution of bulk density of an Inceptisol pedon with Taiwan clay in Southern Taiwan. Copyright by Profs. Hseu and Chen

Fig. 10.4 Saline accumulation on the soil surface after flooding of seawater in Southern Taiwan. Copyright by Profs. Hseu and Chen

deficient in some Inceptisols, Mollisols and Entisols with high pH or from special parent materials in eastern Taiwan.

10.1.7   Water Stress in Deep Sandy Soils

If the soils of coastal areas in western Taiwan consist of large amounts of sand (>70 %), the soils are generally water stressed in the growing season of crop.

10.1.8   Heavy Metal Contaminated Soils

The major anthropogenic sources of heavy metals in the rural soils of Taiwan are illegal wastewater discharged from industrial plants such as chemical, electroplate, and pigment and swine wastewater (Chen 1992). Four stages of survey project of the contaminated soils in Taiwan were performed by the Environmental Protection Administration of Taiwan (Taiwan EPA) since 1982. The objectives of the survey were to understand the heavy metal concentration in the rural soils of Taiwan. The major results of these stages are illustrated as follows (Taiwan EPA 2012).

- Stage 1 (1983–1987): Each representative survey unit is 1,600 ha. The final report was published by Taiwan EPA in 1987.
- Stage 2 (1987–1991): Each representative survey unit is 100 or 25 ha. The results showed that about 790 ha of rural soils have higher concentrations of metals than the regulations of heavy metals announced by Taiwan EPA in 1991 (As 60, Cr 16, Cd 10, Cu 100, Pb 120, Hg 20, Ni 100, Zn 80 mg kg^{-1}). However, the soils were extracted by 0.1 M HCl for the measurement of metals, except As and Hg which were extracted by aqua regia.
- Stage 3 (1992–2000): The survey lands were selected according to the results of Stage 2 which have relatively high concentration of metals in soil. Each representative survey unit was 1 ha.
- Stage 4 (2000–2008): This stage was conducted following the Soil and Groundwater Pollution Remediation Act (SGWPR Act) announced in 2000. The contaminated control or remedial sites were announced based on the criteria of soil control stand (SCS) in the SGWPR Act.

During 1992–1997, the upper levels of background concentration of surface soil (0–15 cm) were determined by the database of heavy metal content in the rural soils of Taiwan as (mg/kg): As 18 ($n = 9,155$), Cd 0.43 ($n = 9,160$), Cr 2.65 ($n = 9,159$), Cu 34 ($n = 9,149$), Pb 18 ($n = 9,157$), Ni 9.7 ($n = 9,170$), and Zn 51 ($n = 9,135$) (As and Hg are total content and other elements are 0.1 M HCl extractable). Moreover, the database of total content of heavy metal in the rural soils of Taiwan has been conducted since 1998. The upper levels of background total concentration of heavy metals in surface soil (0–15 cm) are estimated as (mg/kg): As 18, Cd 2, Cr 50, Cu 35, Pb 50, Ni 50, and Zn 120, respectively (Chen et al. 2007).

Taiwan EPA reported that the areas of contaminated rural soil, which are higher than the SCS, based on the results of 319 ha rural soil survey in 2002, were 159 ha of Ni-contaminated soils, 148 ha of Cu-contaminated soils, 127 ha for Cr-contaminated soils, 113 ha of Zn-contaminated soils, 17 ha of Cd-contaminated soils, 4 ha of Pb-contaminated soils, and 0.3 ha of Hg-contaminated soils. Moreover, the total areas of contaminated rural soil were about 251 ha (Taiwan EPA 2012).

10.2   Soil Contamination and Remediation

10.2.1   Soil Protection Regulation

In comparison to issues of air and water pollution, contaminated lands have been solely paid clear attention in Taiwan for about three decades. Until now, soil remediation technique and regulatory system are needed to be improved for soil contamination, particularly based on human and ecological health. The government of Taiwan started to take account of continuous hazards associated with soil contamination since the 1980s, due to not only a number of environmental disasters, but also increased the regulation and management approaches of soil pollutants in practice from abroad.

In order to fully address soil contamination in Taiwan, the Soil and Groundwater Pollution Remediation Act (SGWPR Act) was announced in 2000 (Taiwan EPA 2012). There are eight chapters and 57 articles respectively in the SGWPR Act. When the concentrations of contaminants exceed the soil control standard (SCS), this site will be announced as a "control site". The polluter of a control site needs to take necessary actions to avoid in further deterioration of contamination. Control sites assessed to have risk on human health by a tiered approach will be further announced as "remediation site". The remediation site cannot be sold and the polluter must remedy the site in accordance with the SGWPR Act. Heavy metal, petroleum fuel, pesticide, and chlorinated organic solvent are major soil contaminants regulated in the SGWPR Act (Taiwan EPA 2012).

Regarding agricultural production, heavy metal contamination in soils is an increasingly urgent problem in the world, and the cleanup of these soils is still time-consuming and difficult. Unlike organic contaminants, heavy metals are generally immutable, not degradable, and persistent in soils. Soils have a natural capacity to attenuate the bioavailability and the movement of heavy metals through them by means of different mechanisms (precipitation, adsorption process, and redox reactions). The concentrations of heavy metals become too high to allow the soils to limit their potential effects, and thus the metals can be mobilized, resulting in serious contamination of agricultural products and the environment. About 90 % of the total number of control sites was rural soil that has been contaminated with heavy metals in paddy fields.

The total area of paddy rice field is approximately 0.4 million ha, which is one-half of agricultural land in Taiwan. In the paddy fields of Taiwan exist a high potential for contamination with industrial and agricultural activities. On the other hand, water resources for rice production in Taiwan have been contaminated by illegal discharge of industrial and livestock wastewater, so that the paddy soil quality and rice safety were adversely affected by heavy metals contamination. Paddy field is characterized with submergence by water. The irrigated water may be contaminated by the above-mentioned source of contamination. As a result, the paddy soils were not only accumulated with heavy metals, but also the residual metals from the wastewater which were retained by sediments in the irrigation canal system. In the long term, the increase of heavy metal level in the sediments also played as a contamination source of paddy soils in addition to the heavy metals directly transported into the rural soils from wastewater.

10.2.2 Soil Heavy Metal Remediation

Many techniques of soil remediation for heavy metal contamination have been developed in the literature, but some of them were not efficient in time, cost, and environmental compatibility. For agro-environmental sustainability, three remediation techniques were illustrated in this chapter according to the remediation experiences in the contaminated paddy soils of Taiwan, including soil turnover and dilution, in situ stabilization by chemical amendments, and phytoremediation in the field.

10.2.2.1 Soil Attenuation

Soil attenuation is an alternative technology in returning heavy metal-contaminated soils to their original function in agriculture. This method adequately mixes contaminated surface soils (0–30 cm) and clean subsurface soils (30–100 cm) in situ, using mechanical forces to reduce the heavy metal concentrations through soil pedon from surface to 1 m depth. This technology process has the advantage of low cost in time and budget, rapid reduction of total concentration of heavy metal, and minimizing the decrease of crop production function after treatment in comparison with other soil remediation techniques such as acid washing and chelating agent extraction techniques in the soils. Any agricultural soil that is identified to be contaminated in meeting the SCS has to be stopped for cropping. All the currently remediated sites of Taiwan were fallow and dried. Nevertheless, the soil turnover was proposed to perform in dry and fallow seasons of cropping rice during October to next March for avoiding clear changes in moisture from subsurface to surface soil. However, this technology is restricted to a relatively lower level of heavy metal soil concentrations at rural land sites and geology settings (i.e., water table, stone proportion) in Taiwan.

Turnover and dilution method was the most popular remediation strategy for most of the metal contaminated paddy soils in Taiwan since 2000. This technical process eventually reduced the levels of organic carbon and other nutrients in the cultivated layers of paddy soils. Consequently, compost and mineral fertilizer were required to be applied into the treated soils for recovering the soil fertility in crop growth and soil function. Additionally, plow pans (around 30–50 cm depth) in the paddy pedons were simultaneously destroyed by the attenuation practices, and thus rebuilding of plow pan was also requested by landowners for following cultivation of paddy rice.

Spatial distribution of heavy metals showed that the metal concentration tended to be high near the entrance of irrigation water within individual field, while the concentration gradually decreased with increasing distance away from the entrance. Therefore, Liu et al. (2008) applied the soil with relatively low concentration of Zn considered as the source of attenuation by horizontal direction for a field-scale remediation practice of paddy soil with Zn up to 4,000 mg kg^{-1} in Changhua, central Taiwan. Furthermore, the amount of Zn was concentrated within 60 cm depth in the paddy soil, but it decreased with increasing soil depth. The soil below 60 cm was considered as the source of attenuation in the vertical direction. Deep plow and consequently mixing well the two layers can significantly decrease the Zn level to meet the SCS of Zn (600 mg/kg) in Taiwan. However, the potential drawback was that the total mass of Zn was consistent through the soil pedon before and after the treatment, so that the bioavailability of metal is needed to evaluate for crop production.

Chen and Lee (1997) used soil turnover and dilution method to effectively decrease the Cd and Pb concentrations in Taoyuan prefecture, northern Taiwan. The soils in the surface layer (0–20 cm) were mixed with the subsoil (20–40 cm) and then were planted by lettuce, water celery, and Chinese cabbage, respectively. These crops were harvested after planting in the treated soils for 4 weeks. Even the mixture of surface and subsurface soils decreased the soil total Cd and Pb concentrations, the mixed soils were still higher than 2.10 mg Cd/kg and 9.07 mg Pb/kg with 0.1 M HCl extraction because the initial metal concentrations were not significantly different on the surface 60 cm. Soil turnover and dilution method can decrease the Cd and Pb concentrations in the crops (Table 10.2), but their eventual total concentrations were still too high for rice production (Chen 1991).

10.2.2.2 In situ Stabilization by Chemical Amendments

Chemical stabilization was evaluated as one of the most cost-effective soil remediation techniques for heavy metal contaminated sites (Chen et al. 2000). Sorption, ion exchange, and precipitation are principal mechanisms to convert soluble and pre-existing potentially soluble solid phase forms of heavy metals to more biogeochemical stable solid phases, reducing the heavy metal pool for root uptake in soil by naturally occurring or artificial additives such as lime material, phosphate, zeolite, bentonite, clay, hydrous iron and manganese oxides, and organic matter (Shuman

1999; Cheng and Hseu 2002). Therefore, it is valuable to understand the levels and mechanisms of controls on heavy metal solubility by different soil additives for remediation practices.

10.2.2.3 Phytoremediation

Phytoremediation was demonstrated to be a feasible method for treating the contaminated paddy fields, which have large areas and low to medium level of heavy metal concentration (Lai and Chen 2005). However, most hyperaccumulator plant species used to remove the heavy metals had low biomass in shoot part and growth rate. Enhanced agents were applied to increase the bioavailability of metals and also increased the amounts of heavy metal uptake in the plants. The application of EDDS, HEDTA, EDTA, or NTA had significant effects on increasing the concentrations of various heavy metals into the shoots of different plants (Meers et al. 2005). The transfer of Cd and Pb from root to shoot (translocation factor, TF) was increased by treatment of EDTA (Chen et al. 2004).

A large field contaminated site in Changhua, central Taiwan, with an area of 1.3 ha and mixed metals contaminated with Cr, Cu, Ni and Zn, was selected to examine the feasibility of phytoextraction in 2006 in Taiwan (Lai et al. 2009). Based on the results of a pre-experiment at this site, a total of approximately 20,000 seedlings of 12 species were selected from plants of 33 tested plant species to be used at this field scale site. A comparison with the initial metal concentration in 12 plant species before planting demonstrated that most species accumulated significant amounts of Cr, Cu, Ni, and Zn in their shoots after growing in this mixed metals contaminated site for 31 days. Among the 12 plant species, the following accumulated higher concentrations of metals in their shoots: *Garden canna* and *Garden verbena* (45–60 mg Cr/kg), Chinese ixora and Kalanchoe (30 mg Cu/kg), Rainbow pink and sunflower (30 mg Ni/kg), French marigold and sunflower (300–470 mg Zn/kg). The roots of the plants of most of the 12 plant species can accumulate higher concentrations of heavy metals than those of the shoots. Extending the growth time from 1 to 2 months promotes the accumulation of heavy metals. These plant species can accumulate higher concentration of each heavy metal, except Zn, in their roots than in their shoots. The phytoavailability of soil metals declines with time during phytoextraction, increasing the required period. Under ideal

Table 10.2 The accumulated Cd and Pb concentrations (mg/kg) in different crops after planting in the turnover and dilution method treated soil for 4 weeks

Treatment	Water celery		Lettuce		Chinese cabbage	
	Cd	Pb	Cd	Pb	Cd	Pb
Control	23.2	4.11	70.4	4.79	49.0	2.76
After dilution	18.4	2.77	39.6	3.44	34.0	6.94

Copyright by Chen and Lee (1997)

conditions and based on the experimental results of this study, it takes 3–20 year to reduce the current concentrations of Cr, Cu, Ni and Zn to the regulation levels based on the SGWPR Act of Taiwan. The results of the large area field scale experiment also show that planting suitable and high-potential phytoextraction plants in heavy metal-contaminated soil can enable the contaminated sites to recover to their natural condition and generate economic benefit because the plants can be sold.

10.2.3 Future Study Needs and Prospects

Rapid civilization makes soil contamination an inevitable problem and a big challenge to scientists and environmental policy makers. For healthy and sustainable future generations, the soil resource should be protected against a slow and insidious poisoning by contaminants released from industrial and agricultural activities in the world. While there are lots of available literatures on soil pollution, the guidelines established by individual countries worldwide to control the pollution of agricultural soils are not consistent and standardized. These points to the complexity of contaminants behavior in agro-environmental system, various climatic, geologic, and hydrological conditions, and the political and nonscientific factors affecting the establishment of regulations in any country.

Hseu et al. (2010) provide a schematic summary of heavy metals in identification, transport, and remediation surrounding the agro-environmental impact in paddy soils in Taiwan. First, considering the central sequence of processes under the environmental monitoring system and according to the regulation of soil pollutants in the SGWPR Act in Taiwan, the source, fate, and remediation of heavy metals for food safety is a dynamic consequence of the time integrated scientific information. However, site-specific and health-based risk assessments are deemed to be the most reliable and practical approaches in resolving the problem of soil pollution. More scientific evidence from case-specific research, especially from long-term field trials involving all kinds of key conditions and factors are necessary to understand the bioavailability of heavy metals in various soil types after long periods of time and to provide reliable parameters for health-based risk assessments. Eventually, the collection of reliable database on heavy metal concentrations in soils and different crops must be given utmost importance. From the database, a variety of useful information and methodology can be developed toward achieving the ultimate goal of providing safe and high quality food for human beings in the future.

10.3 Soil Quality and Sustainability

10.3.1 Definition of Sustainable Management

The FAO/Netherlands conference on Agriculture and the Environment revised the initial definition of "Sustainable Agricultural Development" by FAO in 1990 and translated it into several basic criteria to measure the sustainability of present agriculture and future trends. These criteria can be listed as follows (FAO 1991):

(1) Meeting the food needs of present and future generations in terms of quantity and quality and the demand for other agricultural products.

(2) Providing enough jobs, securing income, and creating human living and working conditions for all those engaged in agricultural production.

(3) Maintaining, and where possible, enhancing the productive capacity of the natural resources base as a whole and the regenerative capacity of renewable resources, without impairing the function of basic natural cycles and ecological balance, destroying the sociocultural identity of rural communities or contaminating the environment.

(4) Making the agricultural sector more resilient against adverse natural and socioeconomic factors and other risks, and strengthening the self-confidence of rural populations.

According to these criteria, the sustainable management of agricultural soils maintains the soil productivity for future generations in an ecologically, economically, and culturally sustainable system of soil management. Sustainable soil management (SSM) must take a multidisciplinary approach. It is not limited only to soil science. Basically, we can consider three aspects of this management system (Steiner 1996):

(1) *Biophysical aspects* Sustainable soil management must maintain and improve the physical and biological soil conditions for plant production and biodiversity.

(2) *Sociocultural aspects* Sustainable soil management must satisfy the needs of human beings in a socially and culturally appropriate manner at a regional or national level.

(3) *Economic aspects* Sustainable soil management must cover all the costs of individual land users and society.

The concept of sustainable land management (SLM) can be applied on different scales to resolve different issues, while still providing guidance on the scientific standards and protocols to be followed in the evaluation for sustainable development in the future (Dumanski 1997). Based on this, sustainable soil management is the basis of SLM, and SLM is the basis of sustainable development (Dumanski 1997).

Land Quality Indicators (LQIs) are being developed as a means of improving coordination when taking action on land-related issues such as land degradation. Indicators are already in regular use to support decision-making at a national or higher level, but few such indicators are available to monitor changes in the quality of land resources. We need more research into LQIs, including integration of socioeconomic (land management) data with biophysical information in the definition and development of LQIs, and screening data for application at various hierarchical levels. The quality of Taiwan's soil is important to the development of Taiwan, and thus we discussing the way to select indicators of soil quality for degraded or polluted soils, and to achieve sustainable soil management.

10.3.2   Soil Quality Concept

Soil quality can be assessed in terms of the health of the whole soil biological system (Warkentin 1995). Many scientists indicate that any definition of soil quality should consider its function in the ecosystem (Acton and Gregorich 1995; Warkentin 1995; Doran et al. 1996; Johnson et al. 1997). These definitions are based on monitoring of soil quality (Doran and Parkin 1994), in terms of:

(1) *Productivity* The ability of soil to enhance plant and biological productivity.

(2) *Environmental quality* The ability of soil to attenuate environmental contaminants, pathogens, and offsite damage.

(3) *Animal health* The interrelationship between soil quality and plant, animal, and human health.

Therefore, soil quality can be regarded as soil health (Doran et al. 1996). Just as we can assess human health, we can evaluate soil quality and health. Larson and Pierce (1994) proposed that a minimum data set (MDS) of soil parameters should be adopted for assessing the health of world soils, and that standardized methodologies and procedures be established to assess changes in the quality of these factors. These indicators should be useful across a range of ecological and socioeconomic situations (Lal 1994; Doran and Parkin 1996). The indicators should:

(1) *Correlate* well with natural processes in the ecosystem (this also increases their utility in process-oriented modeling).

(2) *Integrate* soil physical, chemical, and biological properties and processes, and serve as basic inputs needed for estimation of soil properties or functions which are more difficult to measure directly.

(3) Be relatively *easy to use* under field conditions, so that both specialists and producers can use them to assess soil quality.

(4) Be *sensitive* to variations in management and climate. The indicators should be sensitive enough to reflect the influence of management and climate on long-term changes in soil quality, but not be so sensitive that they are influenced by short-term weather patterns.

(5) Be the components of *existing soil databases* where possible.

Assessment of soil quality is the basis for assessing sustainable soil management in the twenty-first century. It is particularly difficult to select factors of soil quality for degraded or polluted soils. Dumanski (1994) indicated that appropriate sustainable management would require that a technology have five major pillars of sustainability, namely, it should: (1) be ecological protective, (2) be socially acceptable, (3) be economically productive, (4) be economically viable, and (5) reduce risk. Appropriate indicators are needed to show whether those requirements are being met. Some possible soil variables which may define resource management domains are soil texture, drainage, slope and land form, effective soil depth, water holding capacity, cation exchange capacity, organic carbon, soil pH, salinity or alkalinity, surface stoniness, fertility parameters, and other limited properties (Eswaran et al. 1998). The utility of each variable is determined by several factors, including whether changes can be measured over time, sensitivity of the data to the changes being monitored, relevance of information to the local situation, and statistical techniques which can be employed for processing information.

Doran and Parkin (1994) have developed a list of basic soil properties or indicators for screening soil quality and health. They are as follow:

(1) *Physical indicators* including soil texture, depth of soils, topsoil or rooting, infiltration, soil bulk density, and water holding capacity.

(2) *Chemical indicators* including soil organic matter (OM), or organic carbon and nitrogen, soil pH, electric conductivity, and extractable N, P, and K.

(3) *Biological indicators* including microbial carbon and nitrogen, potential mineralizable nitrogen (anaerobic incubation), soil respiration, water content, and soil temperature.

Harris and Bezdicek (1994) indicated that soil quality indicators might be divided into two major groups, analytical and descriptive. Experts often prefer analytical indicators, while farmers and the public often use descriptive descriptions. Soil contaminants selected as indicators may be those which have an impact on plant, animal and human health, or soil function. Soil quality can be viewed from two perspectives: the degree to which soil function is impaired by contaminants, and the ability of the soil to bind, detoxify, and degrade contaminants.

10.3.2.1 Soil Physical Indicators

Hseu et al. (1999) selected some indicators for the evaluation of the quality of Taiwan's soils. The physical indicators included depth of the A horizon, soil texture classes or contents of clay, silt, and sand %, bulk density, available water content (%), and aggregate stability at a depth of 30 cm (Table 10.3). It is easy to understand that measuring the bulk density, soil texture, and penetration of resistance (or infiltration) can provide useful indices of the state of compactness, and the translocation of water and air and root transmission. Measurements of infiltration rate and hydraulic conductivity are also very useful data, but are often limited because of the wide natural variation that occurs in field soils, and the difficulty and expense of making enough measurements to obtain a reliable average value (Cameron et al. 1998). Measuring the aggregate stability gives valuable data about soil structural degradation, which is often affected by pollution (e.g., sodium) and soil degradation (loss of organic matter). This shows that visual assessment of the soil profile is a very valuable way of assessing the physical condition of the soil, and whether there is a need for soil reclamation or remediation. These physical indicators should include:

(1) *Soil texture* related to porosity, infiltration, and available water content.
(2) *Bulk density* related to infiltration rate and hydraulic conductivity.
(3) *Aggregate stability* related to soil erosion resistance and organic matter content.

10.3.2.2 Soil Chemical Indicators

Hseu et al. (1999) selected some chemical indicators for evaluating the quality of Taiwan soils. The chemical indicators included soil pH, electric conductivity (EC), organic carbon, extractable available N, P, and K, and extractable available trace elements (Cu, Zn, Cd, and Pb) (Table 10.4). Standard soil fertility attributes (soil pH, organic carbon, available N, P, and K) are the most important factors in terms of plant growth, crop production and microbial diversity and function. As we know, these parameters are generally sensitive to soil management. For polluted or degraded soils, these soil fertility indicators are regarded as part of a minimum data set of soil chemical indicators.

10.3.2.3 Soil Biological Indicators

Hseu et al. (1999) also selected some chemical indicators for the evaluation of the quality of Taiwan soils. The chemical indicators include potential mineralization of N, microbial biomass of C, N, and P, soil respiration, the number of earthworms, and crop yield (Table 10.5). Soil biological parameters are potentially early, sensitive indicators of soil degradation and contamination. It follows, then, that the minimum data sets for assessing key soil processes are composed of a number of biological (e.g., microbial biomass, fungal hyphae) and biochemical (e.g., carbohydrate) properties (Cameron et al. 1998). Two of the most useful indicators are microbial biomass and microbial activity. Microbial biomass is a sensitive indicator of a long-term decline in total soil organic matter, but does not seem to be a sensitive indicator of the effects of organic pollutants applied to fields.

10.3.3 The Assessment of Soil Quality

There are no reliable, practical methods for assessing or evaluating soil quality/health, although some research reports have established a conceptual framework for assessing this (Karlen et al. 1997). In this chapter, we use the concept of threshold values to evaluate quality of Taiwan rural soils (Cameron et al. 1998). Cameron et al. (1998) suggested that the dynamics of a soil quality value (Q) can be quantified by measuring the changes in soil quality parameters value over time (dQ/dt). This can be done using a quality control chart in which the soil attribute values are

Table 10.3 Soil physical indicators selected for assessing the soil quality of Taiwan soils

Physical indicator	Unit	Relationships to soil function and production	Sensitivity as indicator[a]
Depth of A horizon	cm	Productivity potential and stability of surface	***
Texture (% of sand, silt, clay)	%	Nutrient relations, water relations, root penetration, available water, physical stability, infiltration	****
Bulk density	Mg/m^3	Water and root relations	***
Available water content (1 m)	%	Water retention and release, ratio available water (30 cm/1 m)	***
Aggregate stability (30 cm)	%	Potential for soil erosion, infiltration of water	****

[a] * Very low sensitivity; ** Low sensitivity; *** Moderate sensitivity; **** High sensitivity; ***** Very high sensitivity
Copyright by Hseu et al. (1998)

Table 10.4 Soil chemical indicators selected for assessing the soil quality of Taiwan soils

Chemical indicator	Unit	Relationships to soil function and production	Sensitivity as indicator[a]
pH	–	Biological and chemical activity, limits for plants and microbial activity	*****
EC	dS/m	Plant and microbial activity, limits for plants and microbial activity	*** (or *****)+
Organic carbon	%	Soil fertility, stability, erosion, modeling, potential of productivity	*****
Available N	mg/kg	Plant available nutrient content and potential loss of N, productivity and environmental quality indicators	*****
Available P	mg/kg	Plant available nutrient content and productivity, environmental quality indicators	*****
Available K	mg/kg	Plant available nutrient content and productivity, environmental quality indicators	***
Available Cd	mg/kg	Toxic level for plant quality and crop quality indicator	** (or *****)+
Available Pb	mg/kg	Toxic level for plant quality and crop quality indicator	** (or *****)+
Available Cu	mg/kg	Toxic level for plant quality and crop quality indicator	** (or ***)+
Available Zn	mg/kg	Toxic level for plant quality and crop quality indicator	** or (***)+

[a] * Very low sensitivity; ** Low sensitivity; *** Moderate sensitivity; **** High sensitivity; ***** Very high sensitivity
+ Depend on different crops
Copyright by Hseu et al. (1998)

Table 10.5 Soil biological indicators selected for assessing the soil quality of Taiwan soils

Biological indicator	Unit	Relationships to soil function and production	Sensitivity as indicator[a]
Potential mineralization N	kg N/ha/30 cm	Microbial potential activity, pools of C and N, management effects of organic matter, relative C and N or CO_2 produced	****
Biomass C	kg C/ha/30 cm	Microbial potential activity, pools of C, management effects of organic matter, relative C or CO_2 produced	****
Biomass N	kg N/ha/30 cm	Microbial potential activity, pools of N, management effects of organic matter, relative N produced	****
Biomass P	kg P/ha/30 cm	Microbial potential activity, pools of P, management effects of organic matter, relative P produced	****
Soil respiration	kg C/ha/d	Relative microbial biomass activity, C loss, input and total C pool	****
No. of earthworm	No./m^3	Relative microbial biomass activity	***
Yield	kg/ha	Plant available nutrient content and quality indicators, potential level for crop growth productivity, environmental	*****

[a] * Very low sensitivity; ** Low sensitivity; *** Moderate sensitivity; **** High sensitivity; ***** Very high sensitivity
Copyright by Hseu et al. (1998)

plotted as a time series. The control chart may have a critical limit (or threshold level, or an upper control limit (UCL) and a lower control limit (LCL)) which represents the tolerances beyond which soil quality or other measures of sustainable soil management should not go (Fig. 10.5).

10.3.4 Approaching a National Level of Sustainable Soil Management

10.3.4.1 Action Level of Sustainable Soil Management

Steiner (1996) indicated that the general conclusion drawn from the experiments and experience gained in particular projects can only be transferred to other sites under specific conditions. The solution of the problems in the degraded soils has to be geared to local needs and developed on the basis of a detailed familiarity with local conditions. Programs are needed to coordinate activities at different levels. Sustainable soil management approaches the efforts to get the sustainable agriculture.

Depending on the problems and potential yield, combined actions must be taken at different levels of intervention at the same time (Steiner 1996). These levels can be listed as follows and simplified as shown in Fig. 10.5:

(1) Plot level
(2) Rural household or farm level
   - Technical solution, economically viable
   - Participatory approaches
   - Accounting for target-group specifics

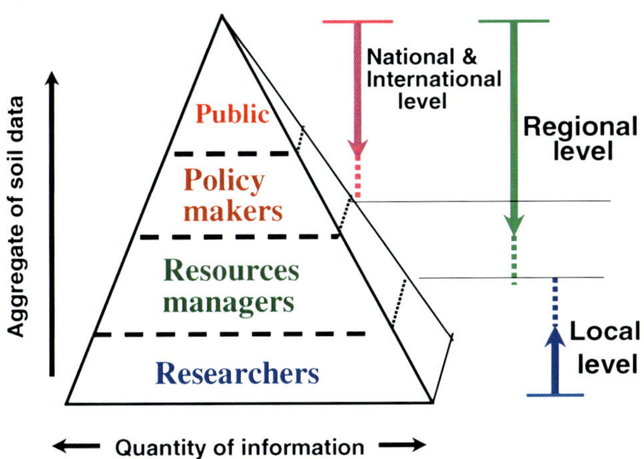

Fig. 10.5 The different communities of people are included in the different levels of operation and proposals for sustainable soil management. Copyright by Profs. Hseu and Chen

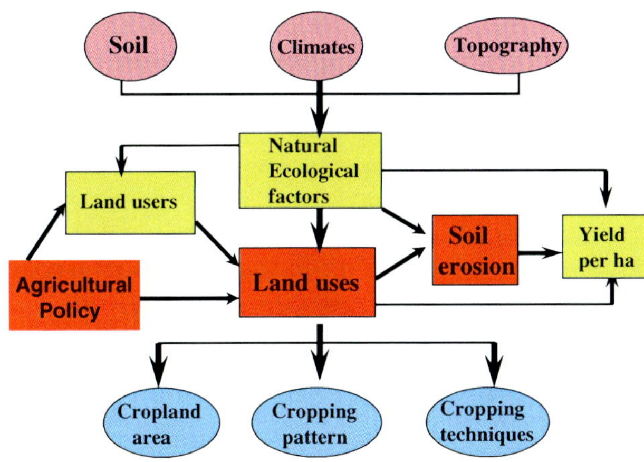

Fig. 10.6 Relationships between soil erosion, land use, and agricultural policy. Redrawn from Maydell (1994)

(3) Village community or watershed level
- Technical solution
- Participatory approaches
- Organizational options

(4) Regional level
- Organizational development (e.g., extension service)
- Land use planning

(5) National level
- National strategies for sustainable soil management
- Agricultural policy (including structure, input supply, and marketing)
- Researches, training, and extension
- Approaches and technical options for sustainable soil management

(6) Supra-regional level
- Collective organization (research institutions)
- Networks for technology transfer and communication

(7) Global level
- Donor coordination
- Trade policy (WTO)
- International research cooperation.

10.3.4.2 Approaches in Agricultural Policy

Maydell (1994) indicated that the research for agricultural policy to promote sustainable soil management must begin by identifying which aspects should be assisted or can be influenced. The chart depicts the relationships between soil degradation, land uses, and agricultural policy (Maydell 1994) (Fig. 10.6). The major factors in these relations are cropland area, cropping patterns, and cropping techniques.

Cropping area Low productivity per ha in most developing countries is the main reason why arable land as a whole has been extended so much in the past and is still being extended today. The only land now left for extension is marginal with low natural soil productivity, unstable soils, and high soil erosion risk. In the past, prices for certain products and subsides have encouraged more land uses. In order to solve these problems in the long term, the most effective way is to raise land productivity by assisting the agricultural research, improving agricultural services, promoting some special crops for farmers, and reducing the input and raising the output.

Cropping patterns Cropping pattern is important for a region. Pricing and active promotion of certain crops can encourage farmers to plant crops that conserve soil (Maydell 1994).

Cropping techniques Production methods have a decisive influence on soil process in the individual farmer. Most methods that conserve soil resources need higher costs. For conserving our soil resources, we need more research on cropping techniques to protect the soil resources.

10.3.4.3 Approaches in Research, Training, and Extension Programs

For research of the project, some questions exist in many countries, such as:

(1) Who defines the needs and aims in research?
(2) Should the research institute or university continue doing this research or should it be the task of extension services or farmers' association?
(3) Who assesses the efficiency of research institute or university and what yardsticks are applied?

From these problems we have, we need cooperation programs among research, technical development, and extension in the national committees, scientific advisory and networks in Taiwan.

10.3.4.4 Approaches and Technical Options for Sustainable Soil Management

The national strategy for sustainable soil management should be established based on the following processes:

(1) Analyze the background and basic data of the degraded soils and polluted soils.

(2) Assemble the components of the effective solution.

(3) Produce a set of tools to establish the national level to meet the needs of farmers and policy makers.

Based on these processes, the approaches for national levels of sustainable soil management can be shown as in Fig. 10.7. The available data on the background of the problem include the causes of soil degradation, current status of soil quality, population and demand of farmers, and agricultural policy. In order to get an effective solution for the problem soils, some components can be considered and proposed. These components include early warming by the soil indicators, prevention of soil degradation, rapid assessment of the problems, assessment on the economic, cultural, and production, or risk assessment for the soil pollutants, and propose a sustainable soil management or control. In order to get a set of valuable tools for needs at national level, we can consider three aspects including policy, tools kits, and communication or education. These three tools are linked to each other for operation of these processes.

Until now, many technical options can be used as components in sustainable soil management system, especially published by FAO and other NGOs (Fig. 10.8). Every soil and water conservation technology must help achieve one of the following goals (Douglas 1994):

(1) Minimize soil erosion (erosion control).

(2) Conserve, improve, and under certain conditions, restore the physical, biological, and chemical properties of the soil (soil fertility and soil structure).

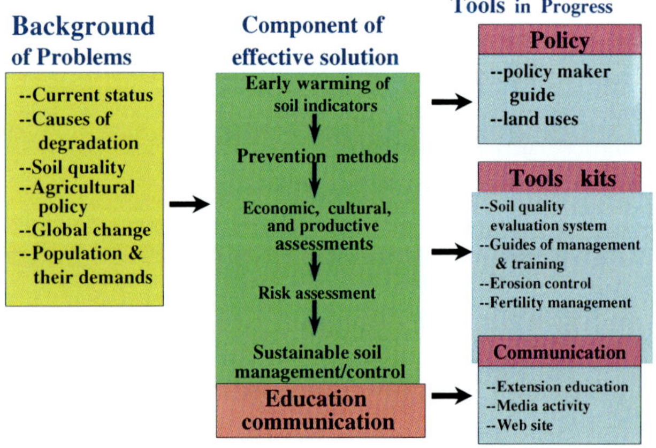

Fig. 10.7 The national strategy of sustainable soil management will assemble the available data on the background of the problem and the net effective solution will then produce a set of tools to be established at national level. Copyright by Profs. Hseu and Chen

Approaches and technical options

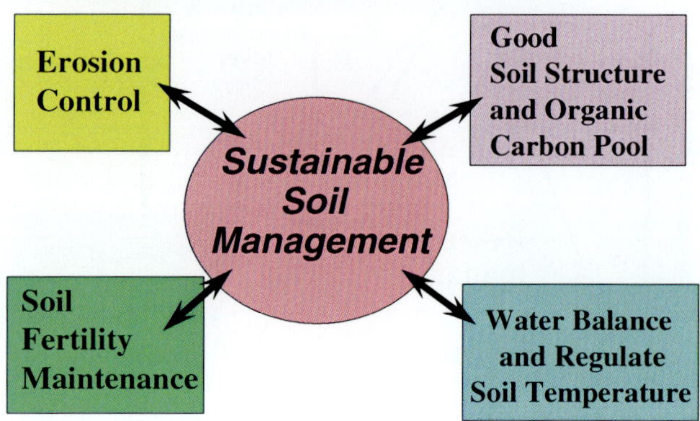

Fig. 10.8 Approaches and technical options can be used as components in sustainable soil management system. Copyright by Profs. Hseu and Chen

(3) Enable the soil to retain water (water balance) and regulate surface run-off.

(4) Regulate soil temperature: higher in uplands, lower in lowlands (temperature control).

Erosion control Vigorous vegetation and rapid ground cover are strongly recommended to avoid soil run-off and soil erosion. The cropping methods include early sowing, cover crop, mixing cropping, high seed density, under seed planting, inter-row cropping, and planting follows. Splash erosion can be controlled by mulching or by leaving harvest crops residues on the soil surface. Rill and gully erosion can be controlled by terracing or other barriers parallel to the slope, such as contour strips planted with different species of grass. Contour plowing and minimum tillage are also effective against different manners of soil erosion. These methods and technologies are not widespread, and they need to be more developed and more extensive to farmers.

Conserving the soil fertility and soil structure As we know, use of harvest crop residues, manure, and composts in the soils is the most recommended way to conserve the soil conditions and keep the soil fertility and soil structure. Another successful method is mulching, where people gather the organic substances (grass, leaves, litters, branches) on non-agricultural fields to avoid soil erosion and to increase the soil fertility in poor soils. The most effective method to maintain the soil fertility, soil structure, and biological activity is providing enough soil organic matter content in the soils, or soil organic carbon pools in the soils (Chen and Hseu 1997). The results indicate that the mean organic carbon content in the surface soils is about 1.9–2.8 % (Guo et al. 2008). The mean organic carbon pools in Taiwan rural soils is less than 8 kg/m^2/m (or less than 5 kg/m^2/50 cm) (Chen and Hseu 1997) (Table 10.6). This organic carbon pool is not enough to maintain good soil structure and crop

Table 10.6 The organic carbon storages (kg/m^2) in three depth intervals of different soil orders in Taiwan

Soil orders	0–30 cm depth			0–50 cm depth			0–100 cm depth		
	Mean	CV	n	Mean	CV	n	Mean	CV	n
Cultivated soils in Taiwan (n = 100)									
Entisols	3.50	46	36	4.97	45	36	7.17	49	36
Inceptisols	3.78	36	35	5.18	32	35	7.68	36	35
Alfisols	3.68	29	8	5.21	30	8	7.85	30	8
Ultisols	3.52	69	10	5.01	38	10	6.53	39	10
Oxisols	2.80	20	3	4.62	15	3	5.41	8	3
Mollisols	8.56	62	3	12.4	45	3	17.3	33	3
Vertisols	8.28	45	3	14.0	30	3	20.7	25	3
Andisols	18.3	–	1	22.5	–	1	27.9	–	1
Histosols	11.1	–	1	18.5	–	1	28.9	–	1

CV the coefficient of variation (%)

N sample number of observation by depth interval

Copyright by Chen and Hseu (1997)

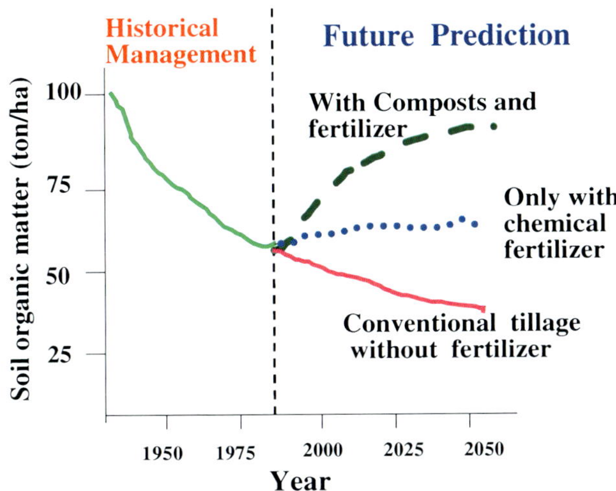

Fig. 10.9 The changes in soil organic matter (tons/ha) in soil under different soil management systems for long-term application of composts or fertilizers. Copyright by Chen et al. (1998a, b)

production. It is necessary the annual application of 20 ton/ha of organic manure or composts to maintain the demands of crop production and provide good soil structure and biodiversity in the soil (Gregorich et al. 1995; Studdert et al. 1997; Chen et al. 1998a, b) (Fig. 10.9).

Soil Water Balance In order to effectively use the limited quantity of irrigation water and precipitation in some areas, some technologies can be applied to achieve the appropriate soil management, such as improving ground cover, conserving organic matter, breaking up (plowing) the soils, roughing the soil surface, building dams, furrows, and contour lines, and terracing the steep slopeland.

References

Acton DF, Gregorich LJ (1995) Understanding soil health. In: Acton DF, Gregorich LJ (eds) The health of our soils: toward sustainable agriculture in Canada. Centre for Land and Biological Resources Research, Research Branch, Agriculture and Agri-Food Canada, Ottawa, pp 5–10

Cameron K, Beare MH, McLaren RP, Di H (1998) Selecting physical, chemical, and biological indicators of soil quality for degraded or polluted soils. In: Proceedings of 16th world congress of soil science. Scientific registration no. 2516. Symposium no. 37. Montpellier, France, 20–26 Aug 1998

Chen ZS (1991) Cadmium and lead contamination of soils near plastic stabilizing materials producing plants in northern Taiwan. Water Air Soil Pollut 57–58:745–754

Chen ZS (1992) Metal contamination of flooded soils, rice plants, and surface waters in Asia. In: Adriano DC (ed) Biogeochemistry of trace metals. Lewis Publishers Inc., Florida, pp 85–107

Chen ZS (1998) Management of contaminated soil remediation programmes. Land Contam Reclam 6:41–56

Chen ZS, Hseu ZY (1997) Total organic carbon pool in soils of Taiwan. Proc Natl Sci Counc ROC Part B Life Sci 21:120–127

Chen ZS, Lee DY (1997) Evaluation of remediation techniques on two cadmium-polluted soils in Taiwan. In: Iskandar IK, Adriano DC (eds) Remediation of soils contaminated with metals. Science Reviews, Northwood, pp 209–223

Cheng SF, Hseu ZY (2002) In-situ immobilization of cadmium and lead by different amendments in two contaminated soils. Water Air Soil Pollut 140:73–84

Chen ZS, Lee DY, Lin CF, Lo SL, Wang YP (1996) Contamination of rural and urban soils in Taiwan. In: Naidu R, Kookuna RS, Oliver DP, Rogers S, McLaughlin MJ (eds) Contaminants and the soil environment in the Australasia-Pacific region, first Australasia-pacific conference on contaminants and soil environment in the Australasia-Pacific region. Adelaide, Australia, Kluwer Academic Publishers, Boston, London, pp 691–709, 18–23 Feb 1996

Chen ZS, Liu JC, Cheng CY (1998a) Acid deposition effects on the dynamic of heavy metals in soils and their biological accumulation in the crops and vegetables in Taiwan. In: Bashkin V, Park SU (eds) Acid deposition and ecosystem sensitivity in east Asia. Nova Science Publication Inc, New York, pp 189–225

Chen ZS, Hseu ZY, Tsai CC (1998b) Total organic carbon pools in Taiwan rural soils and its application in sustainable soil management system. Soil Environ 1:295–306 (in Chinese, with English abstract and tables)

Chen ZS, Lee GJ, Liu JC (2000) The effects of chemical remediation treatments on the extractability and speciation of cadmium and lead in contaminated Soils. Chemosphere 41:235–242

Chen YH, Shen ZG, Li XD (2004) Leaching and uptake of heavy metals by ten different species of plants during an EDTA-assisted phytoextraction process. Chemosphere 57:187–196

Chen ZS, Su SW, Guo HY, Tsai CC, Hseu ZY (2007) New aspects of collaborative research on soil pollution, food safety and soil remediation techniques in Asia. Technical bulletin no. 175. Food and Fertilizer Technology Center (FFTC), Taipei, Taiwan, pp 1–35

Doran JW, Parkin TB (1994) Defining and assessing soil quality. In: Doran JW, Coleman DC, Bezdicek DF, Stewart BA (eds) Defining soil quality for a sustainable environment. Soil Science Society America Special Publication No. 35, Madison, Wisconsin, USA, pp 3–21

Doran JW, Parkin TB (1996) Quantitative indicators of soil quality: a minimum data set. In: Doran JW, Jones AJ (eds) Method for assessing soil quality. Soil Science Society America Special Publication No. 49, Madison, Wisconsin, USA, pp 25–37

Doran JW, Sarrantonio M, Liebig MA (1996) Soil health and sustainability. Adv Agron 56:1–54

Douglas M (1994) Sustainable use of agricultural soils: a review of the prerequisites for success or failure. Development and environment reports no. 11. Group of Development and Environment, Institute of Geography, Berne University, Switzerland

Dumanski J (ed) (1994) Proceedings of the international workshop on sustainable land management for the 21st century, workshop summary, vol 1. Agricultural Institute of Canada, Ottawa

Dumanski J (1997) Criteria and indicators for land quality and sustainable land management. ITC J 1997–3(4):216–222

Eswaran H, Beinroth F, Reich P (1998) Biophysical consideration in developing resource management domains. In Syers JK (ed) Proceedings of conference on resources management domains. Published by International Board for Soil Research and Management (IBSRAM). Proceedings no. 16. Kuala Lumpur, Malaysia, pp 61–78

FAO (Food And Agriculture) (1991) Issues and perspectives in sustainable agriculture and rural development. Main document no. 1. FAO/Netherlands conference an agriculture and environment. S-Hertogenbosch, The Netherlands, FAO, Rome, 15–19 April 1991

Gregorich EG, Angers DA, Campbell CA, Carter MR, Drury CF, Ellert BH, Groenevelt PH, Holmstrom DA, Monreal CM, Rels HW, Voroney RP, Vyn TJ (1995) Changes in soil organic matter. In: Acton DF, Gregorich LJ (eds) The health of our soils: toward sustainable agriculture in Canada. Centre for Land and Biological Resources Research, Research Branch, Agriculture and Agri-Food, Canada, pp 41–50

Guo HY, Liu TS, Chu CL, Chiang CF (2008) Development of soil information system and its application in Taiwan. In: Taniyama I (ed) Monsoon Asia Agro-Environmental Research Consortium (MARCO) workshop: a new approach to soil information system for natural resources management in Asian Countries. Organized by National Institute for the Agro-Environmental Sciences (NIAES) and Food and Fertilizer Technology Center (FFTC). Tsukuba international congress, Japan, pp 141–157, 14–15 Oct 2008

Harris RF, Bezdicek DF (1994) Descriptive aspects of soil quality/health. In: Doran JW, Coleman DC, Bexdicek DF, Stewart BA (eds) Defining soil quality for a sustainable environment. Soil Science Society of America Special Publication No. 35. Madison, Wisconsin, USA, pp 23–35

Hseu ZY, Chen ZS, Tsai CC (1999) Selected indicators and conceptual framework for assessment methods of soil quality in arable soils of Taiwan. Soil Environ 2:77–88 (in Chinese, with English abstract and tables)

Hseu ZY, Su SW, Lai HY, Guo HY, Chen TC, Chen ZS (2010) Remediation techniques and heavy metals uptake by different rice varieties in metals-contaminated soils of Taiwan: new aspects for food safety regulation and sustainable agriculture. Soil Sci Plant Nutr 56:31–52

IUSS Working Group WRB (2006) World reference base for soil resources 2006. World Soil Resources Reports 103. FAO, Rome

Johnson DL, Ambrosce SH, Bassett TJ, Bowen ML, Crummey DE, Isaaxson JS, Johnson DN, Lamb P, Saul M, Winter-Nelson AE (1997) Meanings of environmental terms. J Environ Qual 26:581–589

Karlen DL, Mausbach MJ, Doran JW, Cline RG, Harres RF, Schuman GE (1997) Soil quality: a concept, definition, and framework for evaluation. Soil Sci Soc Am J 61:4–10

Lai HY, Chen ZS (2005) The EDTA effect on phytoextraction of single and combined metals-contaminated soils using rainbow pink (*Dianthus chinensis*). Chemosphere 60:1062–1071

Lai HY, Juang KW, Chen ZS (2009) A large area experiment to study the uptake of metals by twelve plants growing in combined metals-contaminated soils. Int J Phytorem 11:235–250

Lal R (1994) Data analysis and interpretation. In: Lal R (ed) Methods and guidelines for assessing sustainable use of soil and water resources in the tropics. Soil management support services technical. Monograph No. 21. SMSS/SCS/USDA, Washington, DC, pp 59–64

Larson WE, Pierce FJ (1994) The dynamics of soil quality as a measure of sustainable management. In: Doran JW, Coleman DC, Bezdicek DF, Stewart BA (eds) Defining soil quality for a sustainable environment. Soil Science Society of America Special Publication No. 35. Madison, Wisconsin, USA, pp 37–51

Liu YH, Yeh IL, Chen TC (2008) Risk assessment: heavy metals-contaminated soil before and after treatment with an attenuation method. In: Hseu ZY, Chao HR, Chen TC (eds) Proceedings of 2008 international symposium of soil heavy metal pollution and remediation, Pingtung, Taiwan, pp 65–75, 10 April 2008

Maydell OV (1994) Agrarpolitische ansatze zur erhaltung von boden-ressourcen in entwicklungslandern. Landwirtschaft und umwelt. Schriften zur umweltokonomik, Band 9. Wissenschaftscerlag Vauk, Kiel

Meers E, Lamsal S, Vervaeke P, Hopgood M, Lust N, Tack FMG (2005) Availability of heavy metals for uptake by *Salix viminalis* on a moderately contaminated dredged sediment disposal site. Environ Pollut 137:354–364

Shuman LM (1999) Organic waste amendments effect on zinc fractions of two soils. J Environ Qual 28:1442–1447

Soil Survey Staff (2010) Keys to soil taxonomy, 12th edn. NRCS, Washington, DC

Steiner KG (1996) Causes of soil degradation and development approaches to sustainable soil management (English version by Richard Williams). GTZ, Margraf Verlag, Weilersheim

Studdert GA, Echeverria HE, Casanovas EM (1997) Crop-pasture rotation for sustaining the quality and productivity of a Typic Argiudoll. Soil Sci Soc Am J 61:1466–1472

Taiwan EPA (2012) Soil remedial sites of Taiwan. http://sgw.epa.gov.tw/public/En/Default.aspx?Item=r_sites.html. Accessed 8 Dec 2012

Warkentin BP (1995) The changing concepts of soil quality. J Soil Water Conserv 50:226–228

Index

Z.-S. Chen et al., *The Soils of Taiwan*, World Soils Book Series,
DOI 10.1007/978-94-017-9726-9, © Springer Science+Business Media Dordrecht 2015